The Politics of Parametricism

The Politics of Parametricism

Digital Technologies in Architecture

Edited by

Matthew Poole

and

Manuel Shvartzberg

Bloomsbury Academic
An imprint of Bloomsbury Publishing Plc

B L O O M S B U R Y
LONDON • NEW DELHI • NEW YORK • SYDNEY

Bloomsbury Academic

An imprint of Bloomsbury Publishing Plc

50 Bedford Square   1385 Broadway
London   New York
WC1B 3DP   NY 10018
UK   USA

www.bloomsbury.com

BLOOMSBURY and the Diana logo are trademarks of Bloomsbury Publishing Plc

First published 2015

© Selection and Editorial Material: Matthew Poole and Manuel Shvartzberg, 2015
© Individual Chapters: Their Authors, 2015

Matthew Poole and Manuel Shvartzberg have asserted their rights under the Copyright, Designs and Patents Act, 1988, to be identified as Authors of this work.

All rights reserved. No part of this publication may be reproduced or transmitted in any form or by any means, electronic or mechanical, including photocopying, recording, or any information storage or retrieval system, without prior permission in writing from the publishers.

No responsibility for loss caused to any individual or organization acting on or refraining from action as a result of the material in this publication can be accepted by Bloomsbury or the author.

British Library Cataloguing-in-Publication Data
A catalogue record for this book is available from the British Library.

ISBN: HB: 978-1-4725-8166-2
        PB: 978-1-4725-8165-5
        ePDF: 978-1-4725-8168-6
        ePub: 978-1-4725-8167-9

Library of Congress Cataloging-in-Publication Data
The politics of parametricism : digital technologies in architecture /
edited by Matthew Poole and Manuel Shvartzberg.
pages cm
Includes bibliographical references and index.
ISBN 978-1-4725-8166-2 (hardback) – ISBN 978-1-4725-8165-5 (paperback)
1. Architectural design – Data processing. 2. Architecture – Computer-aided design
3. Architecture – Technological innovations. I. Poole, Matthew, editor. II. Shvartzerg, Manuel, editor.
NA2728.P65 2015
720.1'05 – dc23
2015008299

Typeset by Integra Software Services Pvt. Ltd.
Printed and bound in India

Contents

	List of illustrations	vii
	List of contributors	ix
	Acknowledgments	xv
1	Introduction *Matthew Poole and Manuel Shvartzberg*	1
2	The historical pertinence of parametricism and the prospect of a free market urban order *Patrik Schumacher*	19
3	On numbers, more or less *Reinhold Martin*	45
4	There is no such thing as political architecture. There is no such thing as digital architecture *Neil Leach*	58
5	Parametricist architecture would be a good idea *Benjamin H. Bratton*	79
6	*Play Turtle, Do It Yourself.* Flocks, swarms, schools, and the architectural-political imaginary *Manuel Shvartzberg*	94
7	Breeding ideology: Parametricism and biological architecture *Christina Cogdell*	123
8	Speculation, presumption, and assumption: The ideology of algebraic-to-parametric workspace *Matthew Poole*	138

9	Undelete: Recreating censored archives *Laura Kurgan and Dan Taeyoung*	**159**
10	Disputing calculations in architecture: Notes for a pragmatic reframing of parametricism and architecture *Andrés Jaque*	**168**
11	Parametric schizophrenia *Peggy Deamer*	**178**
12	The architecture of neoliberalism *Teddy Cruz*	**189**
13	Parameter value *Phillip G. Bernstein*	**200**
14	Spinoza's geometric and ecological ratios *Peg Rawes*	**213**
	Bibliography	**231**
	Index	**245**

List of illustrations

2.1	Epochal alignments of styles	24
2.2	Medieval town	25
2.3	Palmanova, Renaissance	25
2.4	The Palace of Versailles, the Grand Trianon, ca 1668	26
2.5	Le Corbusier, Ville Radieuse, 1924	27
2.6	Zaha Hadid Architects, Istanbul Master Plan, 2007	28
2.7	The simultaneous enhancement of freedom and order: inversion of architecture's entropy law	40
2.8	Parametricism: Complex variegated order via multiauthor coherence, Studio Hadid, Yale University, 2013	41
4.1	House of the People, Bucharest	59
4.2	Krysztof Wodiczko, projected image of a swastika on South Africa House, London, 1985	65
4.3	Elevation, Plan and Section of Jeremy Bentham's Panopticon penitentiary	66
4.4	Gehry and Partners, Guggenheim Museum, Bilbao, 1997	72
4.5	Gehry and Partners, Disney Hall, Los Angeles, 2003	72
4.6	West Coast Pavilion, Beijing, 2006	74
4.7	Computerized technologies used to produce conventional architectural components	75
6.1	Flock simulation by Craig W. Reynolds, 1987	94
6.2	Reynolds's boid flock circumventing obstacles	99
6.3	Seymour Papert with robotic Logo Turtle, c. 1970	104
6.4	The Architecture Machine Group's SEEK, at the "Software" exhibition, 1970	109
6.5	Nicholas Negroponte's diagram of media technologies convergence, c. 1978	111

7.1	Foreign Office Architects, Classification System of FOA's offspring in *Phylogenesis: FOA's Ark*	124
7.2	Raymond Loewy, Evolution Charts, 1934	126
7.3	*Victimless Leather*, on display at the Museum of Modern Art, New York, in *Design and the Elastic Mind* (2007–2008)	128
7.4	John Frazer with Peter Graham, "Genetic Algorithms and the Evolution of Form"	130
9.1	Python script detail, Jumping the Great Firewall, Advanced Data Visualization Project, GSAPP, Columbia University	161
9.2	Visual interface, Jumping the Great Firewall, Advanced Data Visualization Project, GSAPP, Columbia University	162
10.1	Reconstruction of the urbanism of a Mouride extended family distributed between Paris, Madrid, and Touba	169
10.2	Reconstruction of the urbanism of a family group composed by a mother, her son, and a number of humans and devices distributed between London and Valdemoro	174
11.1	Peggy Deamer, Chart 1	180
11.2	Peggy Deamer, Chart 2	181
11.3	Peggy Deamer, Chart 3	183
11.4	Peggy Deamer, Chart 4	184
11.5	Peggy Deamer, Chart 5	185

List of contributors

Phillip G. Bernstein is a vice-president at Autodesk, a leading provider of digital design, engineering and entertainment software, where he leads Strategic Industry Relations and is responsible for setting the company's future vision and strategy for technology as well as cultivating and sustaining the firm's relationships with strategic industry leaders and associations. An experienced architect, Bernstein was formerly with Pelli Clarke Pelli Architects, where he managed many of the firm's most complex commissions. He teaches professional practice at the Yale School of Architecture, where he received both his BA and his MArch. He is co-editor, with Peggy Deamer, of *Building (In) The Future: Recasting Labor in Architecture* (Princeton: Princeton Architectural Press, 2010), and, also with Peggy Deamer, co-editor of *BIM In Academia* (New Haven, CT: Yale School of Architecture Press, 2011). He is a senior fellow of the Design Futures Council and former chair of the AIA National Contract Documents Committee.

Benjamin H. Bratton is a theorist whose work spans philosophy, art and design. At the University of California, San Diego, he is Associate Professor of Visual Arts, as well as Director of The UCSD Design Theory and Research Platform and D:GP (The Center for Design and Geopolitics). His research is situated at the intersections of contemporary social and political theory, computational media and infrastructure, architectural and urban design problems, and the politics of synthetic ecologies and biologies. His current work focuses on the political geography of cloud computing, massively granular universal addressing systems, and alternate models of ecological governance. His next book, *The Stack: On Software and Sovereignty*, is forthcoming from MIT Press.

Christina Cogdell is a cultural historian who is a chancellor's fellow and Associate Professor in the Department of Design at the University of California at Davis. Her research investigates the intersection of popular scientific ideas and cultural production, in particular art, architecture and design. She is the author of *Eugenic Design: Streamlining America in the 1930s* (Philadelphia, PA: University of Pennsylvania Press, 2004 and 2010), winner of the 2006 Edelstein Prize for outstanding book on the history of technology, and co-editor with Susan Currell of the anthology *Popular Eugenics: National Efficiency and American*

Mass Culture in the 1930s (Athens, OH: Ohio University Press, 2006). Her work has been included in *Visual Culture and Evolution* (Baltimore, MD: The Centre for Art, Design & Visual Culture UMBC, 2012), *Art, Sex, and Eugenics*, edited by Fae Brauer & Anthea Callen (London: Ashgate Press, 2008), and published in *Boom: A Journal of California, American Art, Design and Culture, Volume, Design Issues, and, American Quarterly*. She is currently writing her second monograph on today's "generative architecture" in relation to recent scientific theories of self-organization, emergence and the evolution of complex adaptive systems.

Teddy Cruz established his research-based architecture practice in San Diego, California, in 2000. He has been recognized internationally for his urban research of the Tijuana-San Diego border, and in collaboration with community-based nonprofit organizations, such as Casa Familiar for advancing border immigrant neighborhoods as sites of cultural production, from which to rethink urban policy and propose new models of inclusive housing and civic infrastructure. In 2008, he was selected to represent the United States in the Venice Architecture Biennial, and in 2011 he was a recipient of the Ford Foundation Visionaries Award, the Global Award for Sustainable Architecture, and was named one of the fifty most influential designers in America by *Fast Company Magazine*. He is currently a professor in public culture and urbanism in the Visual Arts Department at the University of California, San Diego, and the co-founder of the Center for Urban Ecologies.

Peggy Deamer is Assistant Dean and Professor of architecture at Yale University. She is a principal in the firm of Deamer Studio. Articles by her have appeared in *Assemblage, Praxis, Perspecta, Architecture and Psychoanalysis* and *Harvard Design Magazine*, among others journals and anthologies. The work of her firm has appeared in *Dwell, The New York Times, Architectural Record* and *House and Garden*, among others. She is the editor of *The Millennium House* (New York: Monacelli Press, 2004) and *Architecture and Capitalism: 1845 to the Present* (London/New York: Routledge, 2013), and co-editor of *Re-Reading Perspecta: The First Fifty Years of the Yale Architecture Journal* (Cambridge, MA: MIT Press, 2005); *Building in the Future: Recasting Architectural Labor* (Princeton: Princeton Architectural Press, 2010); and *BIM in Academia* (New Haven, CT: Yale School of Architecture Press, 2011).

Andrés Jaque is an architect whose work explores the role that architecture plays in the making of societies. In 2003, he founded the OPI (Office for Political Innovation) in Madrid, Spain, a transdisciplinary agency engaged with the making of an ordinary urbanism out of the association of heterogeneous architectural fragments. He has lectured at a number of universities around the world, including Berlage Institute, Columbia University GSAPP, Princeton University, Bezalel Academy, Universidad Javeriana de Bogota and the Instituto Politecnico di Milano, among others. His work has been exhibited at the Schweizerisches Architekturmuseum in Basel, the Instituto Valenciano de Arte Moderno (IVAM), the Biennale di Venezia, where in 2014 OPI won the coveted Silver Lion award, and at the Museum of Modern Art (MoMA) in New York City. He is also the author of other works including Teddy House, Vigo, 2003/2005; Mousse City, Stavanger, 2003; Peace Foam City, Ceuta, 2005; Skin Gardens, Barcelona, 2006; the Museo Postal de Bogotá, Bogotá, 2007; Rolling House for the Rolling Society, Barcelona, 2009; the House in Never Never Land, Ibiza, 2009; the ESCARAVOX, Madrid, 2012; and, Hänsel and Gretel's Arenas, Madrid, 2013. He is currently teaching at GSAPP, Columbia University, New York.

Laura Kurgan is Associate Professor of architecture at the Graduate School of Architecture Planning and Preservation at Columbia University, New York, where she directs the Visual Studies curriculum, the Spatial Information Design Lab and is co-director of the Advanced Data Visualization Project. She is the author of *Close Up at a Distance: Mapping, Technology, and Politics* (New York: Zone Books, 2013). Her work explores things ranging from digital mapping technologies to the ethics and politics of mapping, building intelligence, and the art, science and visualization of big and small data. Her work has appeared at the Cartier Foundation in Paris, the Venice Architecture Biennale, the Whitney Altria, MACBA Barcelona, the ZKM in Karlsruhe, and the Museum of Modern Art, New York. She was the winner of the United States Artists Rockefeller Fellowship in 2009.

Neil Leach is an architect and theorist. He is currently a professor at the University of Southern California and has also taught at the AA, London; Columbia GSAPP, New York; Cornell University, Ithaca, NY; DIA, Dessau, Germany; IaaC, Barcelona, Spain; and SCI-Arc, Los Angeles, California. He is the author, editor and translator of twenty-four books, including *Rethinking Architecture* (London/New York: Routledge, 1997); *The Anaesthetics of Architecture* (Cambridge, MA: MIT Press, 1999); *Designing for a Digital World, Digital Tectonics* (London: Academy Press, 2004); *Digital Cities* (New York: Wiley Press, 2009); *Machinic Processes* (Beijing: China Architecture and Building Press, 2010); *Swarm Intelligence* (LATP, forthcoming); *Scripting the Future* (Shanghai: Tongji University Press, 2012); and *Fabricating the Future and Camouflage* (Shanghai: Tongji University Press, 2012). He has been co-curator of a series of international exhibitions, including the Architecture Biennial Beijing. He is currently a

NASA Innovative Advanced Concepts Fellow working on robotic fabrication technologies for the Moon and Mars.

Reinhold Martin is Professor of architecture in the Graduate School of Architecture, Planning, and Preservation at Columbia University, New York, where he directs the PhD program in architecture and the Temple Hoyne Buell Center for the Study of American Architecture. He is a member of Columbia's Institute for Comparative Literature and Society as well as the Committee on Global Thought. Martin is a founding co-editor of the journal *Grey Room*, and has published widely on the history and theory of modern and contemporary architecture. He is the author of *The Organizational Complex: Architecture, Media, and Corporate Space* (Cambridge, MA: MIT Press, 2003) and *Utopia's Ghost: Architecture and Postmodernism, Again* (Minneapolis: University of Minnesota Press, 2010), as well as the co-author, with Kadambari Baxi, of *Multi-National City: Architectural Itineraries* (Barcelona: Actar Press, 2007). Currently, Martin is working on two books: a history of the nineteenth-century American university as a media complex and a study of the contemporary city at the intersection of aesthetics and politics.

Matthew Poole is a freelance curator and curatorial theorist. His curatorial projects and writing explore the contradictions of neoliberal politics and how they are transforming contemporary art, curatorial practices, the built environment and the political currencies of culture more generally. Previously, Poole was the director of the Centre for Curatorial Studies, in the School of Philosophy and Art History at the University of Essex, UK. There he taught classes on the politics of aesthetics, museology, contemporary curatorial discourses, and contemporary art history and theory. He has curated exhibitions internationally, produced many symposia and conferences and has published on the topics described above. In November 2013, he co-organized, with Manuel Shvartzberg, the conference The Politics of Parametricism: Digital Technologies and the Future(s) of Sociality, at REDCAT (The Roy and Edna Disney CalArts Theatre), in Los Angeles, California, from which this book developed.

Peg Rawes is an architectural historian and theorist, holding the post of senior lecturer and program leader of the Masters in Architectural History at the Bartlett School of Architecture, UCL, London, UK. In both Rawes's teaching and research her work focuses on how aesthetic, material, technological, biopolitical and ecological theories inform contemporary architectural thinking and practice. Rawes has lectured across the world at universities and other public venues, including KTH Stockholm; University of Technology, Sydney; TU Delft; University of Tasmania, Launceston; Tongji University, Shanghai; Hofstra University, USA; University of Dundee; Goldsmiths College, London; Royal Academy, London; Tate Britain, London; and at The Hayward, Serpentine and Whitechapel Galleries, London, among others. In 2013, she was a member of the RIBA Presidents Medals Dissertation Judging Panel. Rawes's recent publications include *Relational Architectural Ecologies*, ed. (London/New York: Routledge, 2013) and *Poetic Biopolitics: Practices of Relation in Architecture and the Arts* (forthcoming 2015, co-editor)—and she is also currently conducting research into "equalities of wellbeing and housing," with Dr Beth Lord (Philosophy, University of Aberdeen), www.equalitiesofwellbeing.co.uk.

Patrik Schumacher is partner at Zaha Hadid Architects and founding director at the AA Design Research Lab, London. He joined Zaha Hadid in 1988 and has since been the co-author of many key projects, including MAXXI—the National Italian Museum for Art and Architecture of the 21st Century in Rome. In 1996, he founded the Design Research Laboratory with Brett Steele, at the Architectural Association in London, and continues to teach in the program. Since 2000, Schumacher is also guest professor at the University of Applied Arts in Vienna. In 2010, he won the Royal Institute of British Architects' Stirling Prize for excellence in architecture. In 2010 and 2012, he published the two volumes of his theoretical opus magnum *The Autopoiesis of Architecture* (London: Wiley Press, 2010 and 2012). His lectures and essays in architectural theory are available at www.patrikschumacher.com.

Manuel Shvartzberg is an architect and writer. He has worked for, among others, OMA/Rem Koolhaas, and was project architect for David Chipperfield Architects in London, where he led a number of international projects between 2006 and 2012. In 2008, he co-founded the experimental practice Hunter & Gatherer, dedicated to pursuing speculative projects on contemporary art, architecture and culture. He has published and exhibited his work internationally and has taught at various institutions, including CalArts and the University of Southern California. Shvartzberg is currently based in New York City, where he is a candidate in the PhD in architecture program and graduate fellow of the Institute for Comparative Literature and Society, both at Columbia University. In 2014, he was part of the team representing the US pavilion at the Venice Architecture Biennale. In November 2013, he co-organized, with Matthew Poole, the conference The Politics of Parametricism: Digital Technologies and the Future(s) of Sociality, at REDCAT (The Roy and Edna Disney CalArts Theatre), in Los Angeles, California, from which this book developed.

Dan Taeyoung is a research associate for Data Visualization at SIDL (Spatial Information Design Lab) in the Graduate School of Architecture, Planning and Preservation (GSAPP) at Columbia University. His work explores the intersection of architecture, politics, technology and community. Taeyoung is also an assistant professor at Columbia's GSAPP, and is designer/co-founder of Troupe Operations.

Acknowledgments

We would like to begin by thanking all of the contributors to this volume—Phillip G. Bernstein, Benjamin H. Bratton, Christina Cogdell, Peggy Deamer, Teddy Cruz, Andrés Jaque, Laura Kurgan and Dan Taeyoung, Neil Leach, Reinhold Martin, Peg Rawes, and Patrik Schumacher—as much for their thoughtful and often provocative interjections into this debate as for their patience, generosity, and hard work on this project.

This book follows the conference "The Politics of Parametricism: Digital Technologies and the Future(s) of Sociality" held at the Roy and Edna Disney CalArts Theatre (REDCAT) in Los Angeles, on November 15 and 16, 2013. We are grateful to the MA Aesthetics & Politics program at CalArts for their hosting of and support in producing the conference under the banner of The School of Critical Studies at CalArts, especially Arne De Boever, Alecia Menzano, Seth Blake, and Amanda Beech. We are also grateful to the team from The Gallery at REDCAT for assisting with the logistics of the conference, including Ruth Estevez, gallery director/curator, and Sohrab Mohebbi, assistant curator. Thanks also to Danielle Dean for her continued encouragement and support.

We are especially grateful to Autodesk for sponsoring the event making these conversations possible in a public forum that opens the debate to wider critical examination. This book would not have been possible without Autodesk's generous support of the conference.

We are also very grateful to Patrik Schumacher and Zaha Hadid Architects for supplying the cover image for this book.

Finally, we would also like to thank the team at Bloomsbury for making this book possible, especially Molly Beck and James Thompson.

Excerpt from BLUE MARS by Kim Stanley Robinson, copyright © 1996 by Kim Stanley Robinson. Used by permission of Bantam Books, an imprint of Random House, a division of Penguin Random House LLC. All rights reserved.

Chapter 1

Introduction
Matthew Poole and Manuel Shvartzberg

We are in the epistemology business.
—Phillip G. Bernstein, vice president, Autodesk, speaking at the Politics of Parametricism conference, November 16, 2013.

Since 2008, "parametricism" has been a recurrent and much-debated theme within architectural discourse. Originally coined by Patrik Schumacher—who takes his work with Zaha Hadid Architects to be the ultimate representation of the term[1]—"parametricism" refers to a type of design process characterized by the interrelation of design variables (or, parameters) through computational tools and techniques; a definition that allows it to encompass the work of other well-known figures and firms as well as emerging practitioners in contemporary architecture and design. Beyond this very general technical definition, however, parametricism has also accrued currency to refer to a whole variety of ideas that animate design culture today, from those concerned primarily with aesthetic questions, to others that are more philosophical, and yet others with strong political agendas. This anthology sets out to explore each of these associations of parametricism and its multiple uses with the explicit aim to shed some light on the political implications in each case. It is our strong rejection of parametricism's most commonly assumed political credo—generally identifiable as liberal democratic "late capitalism," and to which Schumacher's "parametricism" is theoretically welded—that provides the impetus for this book and determines our conscious use and reproduction of the term. If we are using the word "parametricism," it is not to validate it indiscriminately, but rather to open it up to serious critical debate.

Despite its somewhat contested and polemical definition, however, parametricism's currency in architectural culture can be discerned through a variety of channels. It is highly apparent in the curricula of architecture schools around the world; it is frequently present in the discipline's trade publications; and it also features prominently in discussions around architecture's relationship to important collateral fields like engineering and construction. Due to this general presence, parametricism appears to have attained enough popularity for it to have become a quasi-universal label that signals (either tacitly or explicitly) the transfiguration of certain digital design processes to the status of a style, while also becoming an increasingly common practical methodology. Its resonance is thus felt as much in the seemingly instrumental domains of practice—for example, through the efforts to deploy new software technology and construction protocols like BIM (Building Information Modeling) across architecture and associated industries—as much as it is in the guise of a new cultural paradigm capable of agglomerating the more diffused fashions, ideas, and forces of the second decade of the twenty-first century: social networks, big data, "cloud" infrastructures, post-Fordist horizons, global and financial technocracies, and so on.

Parametricism may thus present itself as a case study of the relation between technology, ideology, and contemporary history, where the many tropes and drivers of contemporary politics, economics, and culture—from mass customization and mass inequality, to socially networked revolutions and pervasive global surveillance—come together through and within a specific architectural imaginary. Yet, as a technical and aesthetic repertoire with undoubtedly novel and legitimate preoccupations for the discipline, it is also a highly polemical one in its more or less unconscious relations to political parameters.

This anthology is an attempt to question, map, clarify, and magnify what these political parameters of parametricism may be. Caught between the urge for utilitarian-instrumental innovation, on the one hand, and, on the other, a more unapologetically decadent feast over new aesthetic and formal possibilities, the discourse of parametricism seems to have, so far, a rather limited vocabulary to talk about politics; or, paraphrasing Godard, to "talk politically"—a somewhat puzzling limitation, given parametricism's alignment with all manner of intensified power relations, from racialized labor exploitation to the persistent reproduction of a male-centered architecture culture. This may be, in fact, a problem of contemporary architectural discourse at large (or, the broader cultures in which it operates), but, as already highlighted, it is through the parametric imaginary that the contours of today's powers—our Googles in the Clouds, as much as our many financial apartheids—appear to be most forcefully articulated. As such, it seems imperative to ask some serious questions about both the so-called "realist" and "speculative" deployments of parametricism in order to begin to define its implicit or explicit, actual or potential, political dimensions.

This, of course, is no easy task. Many of parametricism's leading proponents vigorously oppose mixing politics and architecture, taking the view that the two should be construed as structurally exclusive domains of thought and action.[2] In this sense, we may venture to say that most parametricists are not necessarily vocal advocates of this political status quo, but rather tacitly accept the subjacent and normative processes of contemporary neoliberal democracies as their natural or inevitable playing field. Whether for actual political conviction or inertial opportunism, for this position there is little or no credibility in utopian, social, antidisciplinarian, or simply noninstrumental forms of architecture, such as some critical, "projective," or poetic spatial practices. This seems to be at least one of the central political "deficits" or interested elisions on the part of parametricism's most ardent discursive proponents.

We might schematize this state of affairs as follows: On one end of this quasi-naturalized political-architectural spectrum lie positions that orthodoxically uphold the refusal to inscribe politics within parametricism's design processes and broader motivations—a refusal that paradoxically makes it strongly political. On the other end of this spectrum, there are those who uphold the symmetrical position, that parametricism is indeed antipolitical, and thus may hold no clues for either understanding or changing the present status of the discourse and society at large. In other words, parametricism, in its current forms, is at a discursive stasis. Thus, beyond these polarized, negative definitions (parametricism being political in either mirrored refusal of politics), we propose exploring the spectrum itself, willing to entertain the possibility that parametricism may, in fact, contain or project other productive definitions for the intersection of architecture with politics. In order to sunder the lack of critical-imaginative potential that has, so far, hampered the possibility of this kind of investigation within architectural discourse, we suggest enriching our political vocabulary for grasping both parametricism's most radical promises and its deep problematics. This anthology thus proposes to contribute to this effort by showcasing a variety of approaches for determining what is at stake—historically, theoretically, and politically—in parametricism's discursive spectrum.

To a certain degree, then, we assume a certain nonspecificity to parametricism itself—or, in other words, we believe the discourse of parametricism to be sufficiently diverse, contradictory, and heterogeneous, *and yet* sufficiently problematic and politically substantial—even "dangerous"—to merit serious critical attention. The form we have given this attention, in these circumstances, is that of a survey that attempts to draw and clarify these various dimensions of the discourse. The positions staked out by our various contributors do not—on the whole—take the entrenched postures outlined above, but rather pose serious questions for the extant theorizing, historicizing, and practical applications of parametricism to complicate and expand its discourse. Our editorial approach has been to encourage this critical engagement through the understanding of the actual conditions and techniques involved in parametricism, rather than through the more removed position of a purely ideologically driven critique. As the selected epigraph for this introduction suggests, we are interested in the conjugation of architecture, design, information, economics, communications, and computational technologies not simply as a means of "managing" and productively exploiting the actions, communications, knowledge, and thought of actors in a social field, but rather as powerful and active practical performative processes of remodeling and rebuilding extant theories and media *of* knowledge—the technical and epistemic substrates of our political realities—via the material and immaterial structures that are constructed by these increasingly overlapping industries. This materialist and posthumanist approach to the complexity of such imbricated socio-technical ecologies provides a framework to address and contest social assemblages, including nonhuman actors.

Through both historico-technical and socio-philosophical critique, we hope to begin to take some of the necessary steps toward mapping a better understanding of this complex topic: sorting what is and what is not "parametricism," as well as what its other particular histories, applications, and potentials might be—in historical, technical, and intellectual terms. This process of discourse-genealogy attempts to address parametricism from important historical vantage points that the discourse itself, as yet, is not in general sufficiently aware of, as well as opening up methodologies of practice and inquiry that strongly problematize parametricism's politics as it currently stands, while not necessarily rejecting all of its attributes, techniques, or modes of operation. In sum, we hope to capture how the current politics of parametricism was *and is* cultivated—and how it could *and should* be different.

What is "parametricism"?

A necessary aspect of this genealogy involves attempting to define more accurately what parametricism *is*, even if only in a summary fashion. Rather than definitively adjudicate on its definition, however, we will attempt to give a general introduction of what it does, and of the heterogeneous terrain it arises from, illustrating where its various lines of contention appear to be currently placed. Crucially, this involves trying to distinguish "parametricism" from the closely associated terms "digital design," "algorithmic design," and perhaps most importantly, "parametrics"/ "parametric design."

As noted above, the word "parametricism" was coined by Patrik Schumacher,[3] who is the strongest proponent for the word itself and for its association with neoliberal politics and economics.[4] We will not dwell on the purely spectacular dimensions of Schumacher's architecture and discourse, which, in their spectacularity, perform the role of commodities in an ever-more competitive global capitalist market for images and status production—a fundamental aspect which has a long and still highly relevant history within Marxian ideology critiques. However, it is important to distinguish Schumacher's "parametricism" from the more diverse field of "parametrics" in architecture. Parametricism's novelty, according to Schumacher, lies in the new formal-procedural repertoire made available by new computational tools and design processes. These formal-procedural changes in the "basic, constituent elements of architecture"[5] signal shifts in the capabilities and ways of thinking of designers: "Contemporary (Parametricist) architects start to conceive of urban fields as swarms of buildings, continuously differentiated, yet coordinated with respect to size, distance and orientation, creating fluid swarm formations."[6]

Beyond this technical-formal definition—which constitutes, significantly, another iteration of the modernist project of purely formal rationalization rather than a departure from it—Schumacher also explicitly attempts to inscribe the term "parametricism" as the signifier of a new aesthetic and cultural paradigm—the new incarnation of the "avant-garde," which explains his use of the suffix *-ism*. Specifically, he considers parametricism to be the "epochal style" of the "post-Fordist network society"[7] that emerges after his so-called "transitional periods" of postmodernism and deconstructivism. In this teleological scheme of styles corresponding to the "grand epochs" of Western civilization, parametricism is, according to Schumacher, poised to become the new hegemonic movement for the twenty-first century, replacing Modernism as the new contemporary "International Style."[8] He thus considers himself and his project—and others like it[9]—to be the transformational avant-garde of the discipline that will usher in this epochal change.

"Parametrics" or "parametric design," by contrast, is not necessarily associated with a style or a movement, but refers to the adoption of a very broad series of techniques of computation in virtual modeling processes. These modeling processes are developed either through the use of custom or proprietary software where the designer uses a preconfigured user interface or "plug-in"—designed to make the manipulation of formal or other parameters easier, more intuitive or more user-friendly—or they are developed through the actual programming of algorithms that are designed to set in motion procedures for interrelating variables, parameters, or specific conditions—formal or otherwise. This second kind of process, more accurately defined as "algorithmic design," is coded through computer "scripting" by generating, inputting, or mediating parameters—hence the common slippage between the terms "algorithmic design" and "parametric design."

More generally, "the parametric" is a semantic pool that has its foundations in mathematics and its main applied uses in statistics, science, and computing, nowadays associated with complex systems analysis or engineering requiring the calculation of large quantities of data processing. Growing out of these underlying disciplinary fields, parametric calculation and analysis offers a tool for the representation and transformation of a variety of conditions: not only abstract or formal, but also applied to environmental, structural, energetic, or even logistical, organizational, financial, and managerial models.

Thus, we consider parametric technology, employed in different ways by Google, NASA, Bloomberg, Skanska, or Gehry Technologies, among others, as too broad and used in too many contradictory ways to constitute a unified style or movement. This is not to deny "the parametric" as a potential cultural condition—as virtual modeling and computation (including the handling of "big data") may be considered paradigmatic of the way in which the digital permeates many aspects of our contemporary societies—but such a condition is too generic, complex, and multifarious to form the scope of an all-encompassing, single theoretical architectural project in the way Schumacher has done by transforming "parametrics" into "parametricism." While Schumacher's own treatises—which seek to fully entrench the term as the definitive moniker for the most relevant and historically conscious type of architecture today and in the future—have become a highly visible and much-debated body of literature on the subject, most designers elude Schumacher's strong nominalism and theories. Rather, they tend to use "parametrics" as an umbrella term for a variety of different practices.[10]

Indeed, a much broader set of practices and definitions would have to be addressed should one look at the history of "parametrics" in design generally, even beyond that which we call the recent history of "digital design" of the last fifty or so years.[11] As many authors have noted, all design is at some level "parametric," as Mark Burry put it: "'Parametric design' is tantamount to a *sine qua non*; what exactly is non-parametric design?"[12] Thus, within this necessarily broad and therefore quite generic discursive landscape, we take the term "parametric design" to refer to a type of virtual modeling that uses parametric calculation where the "shape" and morphology of "forms" (meant here in the broadest possible terms, including social forms) is determined by dynamic and recursive streams of inputs—a capability that can be explored and applied in a large number of extremely different ways.

Within the design fields, the popularization of these different digital design tools was originally driven from two apparently opposite directions. On the one the orchestrated confluence of postwar (and some prewar) sciences, technologies, and obsessions, such as the focus on cognitive–behavioral psychology and the interest in executable modes of codification, whether as pedagogical curricula, computer programs, or urban design protocols. Parametricism, to a certain degree, represents the nth iteration of this type of methodology, even though today it is infused with a different distribution of "sciences," tools, and motivations—from neuro-marketing to real estate, from issues of social justice to environmental collapse, and more.

There is clearly much work to be done in this realm of digital design's multiple "prehistories" beyond the limited set of case studies mentioned above, and while some of the contributors to this volume do engage with this necessary task, more thorough-going critical-historical works recounting and weighing the relative importance of these myriad technical and discursive interventions (for "parametricism" as much as for other strands of contemporary digital design) are patently yet to be written.

However, the very act of recognizing these various "programs" that both "run" and "run on" parametricism, through and beyond its literal procedural functionalities, as well as the recognition of their similarities with earlier techno-ideological paradigms and ideals of "the political," constitutes the starting point for the critical discourse we are attempting here. To this end, as well as the rigors of historical examination, there are also a variety of contemporary methods, approaches, and intellectual frameworks that we consider indispensable for the task—some of which we will now briefly introduce.

The political

You will be wondering at this point what do we mean by "the political" and "politics"? Answering this question is obviously a mammoth task, and while there are many iterations by a myriad of thinkers who have addressed it, here we can only give an extremely partial description. Yet, we believe the relation between parametricism and politics would be best served by considering, at least, the following fundamental fields and authors.

At some level, all of the humanities and social sciences—from comparative literature to anthropology, sociology, etc.—touch on the question of politics, but a common starting point into the discourse is the work of Hannah Arendt, who worked from Ancient Greek, Roman, and Modern intellectual sources to detail the complexities of the aesthetic and performative dimensions of "the political."[23] Reading Arendt would open up other political thinkers and philosophers of note, including classics from Thomas Hobbes to Karl Marx and beyond, with the whole tradition of the Enlightenment in between. On the way, some important theorists of related fields, like political economy (i.e., Adam Smith) and law (i.e., Carl Schmitt), would also make an appearance. Among more recent political philosophers, we would count Antonio Gramsci, Walter Benjamin, Jürgen Habermas, and others, who have inspired the work of thinkers as disparate as Michel Foucault, Jacques Derrida, Claude Lefort, Ernesto Laclau, Gayatri Chakravorty Spivak, Chantal Mouffe, and many more.[24]

Of course, this partial list is but a drop in the ocean of political thought. Yet, in its composition we can observe the historical importance of diverse themes such as ethics, abstraction, or language—and their various disciplinary offspring; law, economics, semiotics, phenomenology, etc. The contemporary political philosopher Jacques Rancière, to give just one example, draws from both poststructuralist and Ancient Greek philosophies to theorize a "logic of politics" that seeks to capture the dynamic of social change as a radically contingent historical process where administration (i.e., economy, law, security) and democracy (social movements' concrete historical struggles for equality) are always at odds with each other. Politics is thus the very friction generated by this encounter.[25] Rancière's theory is elegant for its formal logic, posing a quasi-universal yet indeterminate procedure for democratic politics, articulating an antiessentialist understanding of social actors (i.e., nonidentitarian) through a theory of epistemology *becoming* history: politics as the concrete and symbolic action of the "redistribution of the sensible"—the redefinition of what things mean in the sense of what and how societies' different "parts" get accounted for, or not.

Yet, Ranciere's political theory, like many others from the poststructuralist tradition, rests on a legacy of semiotics whereby social processes are understood mainly as symbolic events, rather than seeing them more fundamentally as material (and thus, technical) traces and forces. In this sense, there is today a renewed interest in the role of materialities as well as techniques and technologies, to think "the political." Media-technical and materialist theories (for instance, all those figured under the umbrella of Science and Technology Studies (STS) best represented by the work of Bruno Latour[26]) thus provide us with an opportunity to theorize "the political" *from within* material and technical discursive formations (such as parametricism) through their particular processes of mediation, rather than as purely symbolic projections by ideological actors and forces.

An important pioneer for this theoretical method is the work of Michel Foucault, who developed an influential approach toward the historical critique of discursive formations. Among his key innovations in this area is the field of inquiry now associated with "biopolitics"—influencing, today, the work of thinkers such as Antonio Negri, Giorgio Agamben, and Judith Butler, among many others.

"Biopolitics" was originally proposed by Foucault as a lens through which to understand the history of governmental theories and techniques—reflecting States' progressive hold over "life" in all its forms: via policies and practices around education, regimes of care and health, etc.—and also provided him with a banner-term to discuss the technical, ideological, and theoretical dimensions of liberalism, an important historical-political actant over the last three centuries.[27]

In liberalism, Foucault finds an articulated technical and ideological program for organizing social life around economic activity. This new control over civil society's "nature," disclosed by the development of "economic sciences" such as statistics, makes liberalism a kind of "governmental naturalism."[28] According to Foucault, the governmental principle of liberalism is to govern according to the market because the market is understood as a site of truth—the natural truth implicit in the relation of exchange between autonomous individuals whose interests cannot be externally anticipated or known, and which therefore only the spontaneous operations of the market can compute. Liberalism thus poses the principle of laissez-faire as the most expedient mechanism for governing—finding the right measure between governing "too much" and "too little" according to what "naturally" works in the market.

The history and theory of liberalism is important for understanding "neoliberalism," a term already mentioned a few times in this introduction, as well as in many chapters of this anthology. Indeed, intensifying liberalism's focus on economic scientificism over the social sphere, neoliberalism, from its early diverse formulations to the more consolidated form we see today, essentially operates by producing a political field that subsumes "political agency," responsibilities and effects of political institutions (such as government bodies, State-run social welfare agencies, workers' unions) into the machinations of economics at various interconnected and "nested" scales (from the personal and local to the national and global) of capitalist economies. That is to say that it subsumes political action into the operations of "the free markets" of trade and exchange between private businesses, corporations, and individual private citizens in order to attempt to increase economic efficiency and sustain growth within national economies across the globe. To achieve this, proponents of neoliberalism argue for economic liberalization via the mechanisms of free trade, privatization of public services, deregulation of financial and industrial working processes and methods, as well as through massive scaling down of government spending at all levels (except, in most cases, for national defense budgets).

The neoliberalism that is referred to most often in this book is that which is most closely allied to the work of the Austrian/British economist Friedrich Hayek

(1899–1992) and the American economist Milton Friedman (1912–2006) of the Chicago School of Economics, which have developed into the global economies that we see today reproducing, reforming, and further intensifying the types of policies that British prime minister Margaret Thatcher and US president Ronald Reagan (both heavily influenced by Hayek and Friedman) introduced and consolidated between the two nations in the 1980s.

However, the term and the concept have complex and complicated histories and iterations—ranging from the most formal and empirical (such as those that generated the highly influential Efficient Markets Hypothesis) to the more classically bourgeois, socially minded variants (such as the "ordoliberalism" of the Freiburg School, which posed a "social market economics"). All of these early incarnations of neoliberalism favored the protection of the tenets of liberalism per se; that is, the "inalienable rights" of the individual to liberty, private property, social and political equality and fraternity within the State; but disagreed on how these were to be achieved within the interrelationship and interoperability of public State institutions and private capitalist markets. In all cases, however, neoliberalism aims to "govern *for* the market, rather than *because* of the market"[29]—a subtle but crucial shift from liberalism highlighted by Foucault. And, of course, the very definition of large and ambiguous concepts like "inalienable rights" or "fraternity" mobilizes radically different political imaginaries and philosophical interpretations, which is why a critical focus on specific techniques and materialities may be of particular use to confront the system that reproduces them.

Displacing questions of ontology and ideology (such as the classical questions of authentic motives or absolute fairness) to analyze, instead, neoliberalisms' various technologies of governance is also historically appropriate. As Foucault describes, the "marketization" of society is in fact historically tied to an engagement with diverse mid-twentieth-century sciences, thus enabling the development of neoliberal rationality as a techno-epistemic project. The theorization of neoliberal economic-governmental programs (like, for instance, behavioral economics) both mobilized and depended on cognitive and behavioral psychology, systems theories, computing, as well as techniques of simulation and control of "the environment" as a whole, such as cybernetics.

In other words, the technologies of user-codification and association, now activated by subjects themselves through the full subsumption of the life process within techno-epistemic capitalist interfaces, makes the discussion of the politics of parametricism particularly relevant in this context. As parametric logics and environments continue to be deployed at ever-greater scales (from the very micro to the very macro) and across disciplines and territories, they more effectively establish fundamental characterizations of identity, citizenship, freedom, and other relational markers of subjectivity, at the same time as they provide for new communicative and experiential possibilities—a new techno-political reality of quasi-invisible, embedded, pervasive systems of environmental modulation and control, for better *and* for worse.

Many of the contributors to this anthology thus refer to key figures dominant in the politico-philosophical interrogations of these conditions, such as Jean-François Lyotard, Gilles Deleuze, Félix Guattari, and Alain Badiou, among others. Lyotard's 1979 book, *The Postmodern Condition: A Report on Knowledge*,[30] is especially important in this context as it outlines the impact of digital computational and networked communications technologies on the changing status and value of knowledge in postindustrialized societies, exploring and speculating upon the shift from knowledge to "information" and the possibilities and ramifications of information as a commodity form in the face of industrial "globalization." Much of Deleuze and Guattari's writing follows such a path also to interrogate the complex interplay of social and market forces upon the changing ecology of epistemology, aesthetics, and ontology in a technologically driven global world. Badiou's body of work intervenes in perhaps the most militant manner in the politico-philosophical debates around technology, capitalism, and neoliberalism, where we see his recasting of ontology as a specifically mathematical function. For the issues outlined in this book, the question of the computation of numbers and the issues that develop due to the increasingly complex technologies of calculation, which are developing in and beyond the fields of architecture and design, must be addressed in order to understand their effects upon all social and political actors and actants and the forces that they exert upon one another in socio-political ecologies, which are more wide reaching and interconnected in their effectivity than ever before.

Within this sphere, the recent work of a younger generation of philosophers and theorists of technology and politics has also been instructive to the issues that are presented here. The Italian *Operaismo* and *Autonomia* theorists, including Paolo Virno, Christian Marazzi, Maurizio Lazzarato, Franco "Bifo" Berardi, and Antonio Negri, among others, present compelling problematizations of the techno-utopianism often associated with post-Fordist labor models and post-Taylorist management structures, both of which are made possible by digital networked communications and digital electronic computation.[31] These writers explore how capitalist activity is now able to evaluate, process, and intervene within not only the semiotics of social interactions but actively embeds within the codes of the communicative, affective, and behavioral aspects of social life. Latterly, this approach has continued through the work of writers associated with "Accelerationism"; a loose grouping of theorists, philosophers, designers, and artists all of whom are dedicated to investigating what they see as the stultifying effects of capitalism and neoliberalism upon the possible future flourishing of digital electronic technology and its potential to assist in a radical reorientation of the social and political foundations of a globally connected world.[32]

These broad areas of inquiry, between technology and social organization, resonate—indirectly, for now—with architecture's role in economies and geographies at large, such as in the way architectural services are embedded in supply chains, the construction industry, the status of design as an "immaterial" service, as well as with theorizations around the possible repurposing of capitalist forms (like corporations) toward more cooperative arrangements. These theoretical and practical experiments, of which parametricism has so far been relatively disengaged, pose new challenges for the discipline as such, opening up design to a broader realm of cultural production that might be variously combined with fabrication, entertainment, community services, learning, curatorship, publishing, academic discourse, and many other activities, both in physical urban space and organized entirely through shared digital infrastructures. Most importantly, and as many of this book's contributors explore, parametrics are likely to radically affect the media of political economies *per se*—that is, the specific techno-cultural relationships between the governed and the governing, through novel entities and new complex dynamic chains of decision-making, logistics and representation. Politicizing these relationships, rather than leaving their fate to the always and already constraining ideological forces of 'markets,' is the central drive of this book and our wider project to interrogate and activate the politics of parametricism.

Indeed, it is the capacity for data and its morphologies to be embedded in all these different realms simultaneously that determines and defines its unique status as the "object" of contemporary political action and thought. Recognizing the deep consequences of seeing data processes, infrastructures, and political representations as integrally composite articulations potentially opens a way beyond techno-utopian naiveté as much as techno-phobic impasses—a move toward hybrid modes of practice and research that might one day shift, as Arjun Appadurai has recently called for, our current and ambivalent "politics of probability" to a brighter "politics of possibility."[33]

Notes

1. "Parametricism as Style—Parametricist Manifesto", by Patrik Schumacher, London, 2008. Presented and discussed at the Dark Side Club1, 11th Architecture Biennale, Venice. Found at http://www.patrikschumacher.com [accessed October 1, 2013]. Zaha

Hadid Architects' notable projects include the Vitra Fire Station in Weil am Rhein (1994), Phaeno Science Center in Wolfsburg (2005), or the MAXXI museum in Rome (2010).

2. See, for instance, the chapter "Is Political Architecture Possible?" in Patrik Schumacher, *The Autopoiesis of Architecture. Vol. II: A New Agenda for Architecture* (Chichester: Wiley, 2012).

3. Among other articles and books, Schumacher's publications include *The Autopoiesis of Architecture* (Chichester: Wiley, 2012); *Latent Utopias* (Wien: Springer, 2002); "Patrik Schumacher on Parametricism—'Let the Style Wars Begin,'" *Architects Journal* Vol. 231, No. 16 (May 6, 2010), pp. 7–48; "Parametricism: A New Global Style for Architecture and Urban Design," *Architectural Design* Vol. 79, No. 4 (2009), pp. 14–23; and many more texts on his personal website http://www.patrikschumacher.com [accessed October 1, 2013].

4. Schumacher's own brand of neoliberalism is perhaps most closely related to right-wing Libertarianism. There is a wealth of volumes on the many geographical and actually implemented variants of "neoliberalism." For a brief critical introductory overview, see Jamie Peck, *Constructions of Neoliberal Reason* (Oxford: Oxford University Press, 2010). See also: Daniel Stedman Jones, *Masters of the Universe: Hayek, Friedman, and the Birth of Neoliberal Politics* (Princeton: Princeton University Press, 2012); David Harvey, *A Brief History of Neoliberalism* (Oxford: Oxford University Press, 2005). Aihwa Ong, *Neoliberalism as Exception: Mutations in Citizenship and Sovereignty* (Durham, NC: Duke University Press, 2006).

5. Schumacher, "Patrik Schumacher on Parametricism," (May 6, 2010).

6. Schumacher, *Autopoiesis, Vol. 1*, p. 423.

7. Schumacher, "Patrik Schumacher on Parametricism," (May 6, 2010).

8. Schumacher, *Autopoiesis, Vol. 1*.

9. Schumacher mentions Greg Lynn Form, UN Studio, and Reiser and Umemoto as practices working along the lines of "parametricism." See, Patrik Schumacher and Peter Eisenman interview, "I Am Trying to Imagine a Radical Free-market Urbanism," in eds. Peter E. Eisenman & Anthony Vidler, *Log 28—Stocktaking* (New York: S.I. Anyone Corporation, 2013), p. 42.

10. See, for instance, eds. Peggy Deamer & Phillip G. Bernstein, *Building (in) the Future: Recasting Labor in Architecture* (New Haven, CT: Yale School of Architecture, 2010); the work of Space Syntax, UCL's spin-off company; the Spatial Information Design Lab (SIDL) and Laura Kurgan's *Close Up at a Distance: Mapping, Technology, and Politics* (Brooklyn, NY: Zone Books, 2013).

11. See, for instance, Mario Carpo, *The Alphabet and the Algorithm* (Cambridge, MA: MIT Press, 2011). Also, Luigi Moretti wrote extensively in the 1940s about "parametric architecture" as the study of architecture systems with the goal of "defining the relationships between the dimensions dependent upon the various parameters." Luigi Moretti, "Ricerca Matematica in Architettura e Urbanisticâ," *Moebius* Vol. IV, No. 1 (1971), pp. 30–53. Republished in, Federico Bucci & Marco Mulazzani, *Luigi Moretti: Works and Writings* (New York: Princeton Architectural Press, 2000), p. 207.

12. Mark Burry, *Scripting Cultures* (Chichester: Wiley Press, 2011), p. 18. Or, as Aish and Woodbury assert: "Parametric modelling is not new: building components have been adapted to context for centuries." Robert Aish & Robert Woodbury, "Multi-level Interaction in Parametric Design," in *Smart Graphics: 5th International Symposium*, eds. Andreas Butz, Brian Fisher, Antonio Krüger, & Patrick Olivier (Frauenwörth Cloister: Springer, 2005), p. 152.
13. See, for instance, Greg Lynn, *Animate Form* (New York: Princeton Architectural Press, 1999).
14. Procedurality: "When you model using parametrics you are programming following similar logic and procedural steps as you would in software programming. You first have to conceptualize what it is you're going to model in advance and its logic. You then program, debug and test all the possible ramifications where the parametric program might fail. In doing so you may over constrain or find that you need to adjust the program or begin programming all over again because you have taken the wrong approach." Rick Smith, *Technical Notes from Experiences and Studies in Using Parametric and BIM Architectural Software*, published March 4, 2007, at http://www.vbtllc.com/images/ VBTTechnicalNotes.pdf [accessed September 25, 2014], p. 2.

    Flexibility: "Changing the order in which modelling and design decisions can be made is both a major feature of and deliberate strategy for parametric design. Indeed, a principal financial argument for parametric modelling is its touted ability to support rapid change late in the design process." Robert Woodbury, "Chapter 1," *Elements of Parametric Design* (Abingdon: Routledge Press, 2010), p. 43.

    Variability: "Initially, a parametric definition was simply a mathematical formula that required values to be substituted for a few parameters in order to generate variations from within a family of entities. Today it is used to imply that the entity once generated can easily be changed." Chris Yessios (founder and CEO of the modeling software FormZ), "Is There More to Come?" in *Architecture in the Digital Age: Design and Manufacturing*, ed. Branko Kolarevic (New York: Spon Press, 2003), p. 263.

    Correlation: "parametric modelling introduces fundamental change: 'marks,' that is, *parts of the design*, relate and change together in a coordinated way." Burry, *Scripting Cultures*, p. 9.

    Interdependency: "The challenge of building a parametric model is to untangle the interdependencies created by different requirements and find a set of rules that is as simple as possible while remaining flexible enough to accommodate every occurring case. In other words: to pinpoint the view to the exact level of abstraction where no important point is lost and no one gets distracted by unnecessary detail." Robert Woodbury, "Chapter 1," p. 11.
15. Fabian Scheurer & Hanno Stehling, "Lost in Parameter Space?" *Architectural Design 81* Vol. 4 (2011), p. 75.
16. Lev Manovich, *Software Takes Command: Extending the Language of New Media* (New York & London: Bloomsbury, 2013); and Carpo, *The Alphabet and the Algorithm*.

17. "(…) one of the more noticeable trends in recent architecture is the turn from metaphysics to immanence. Whereas postmodern architects thought of buildings and cities as fragments of an expansive texture or fabric, contemporary spaces are mostly involuted and introverted. The turn to immanence takes on different forms in current architectural practice. The resulting architecture in no way encodes the design process in the contours of indexical form, but sublimates it into asignifying, involuted worlds for which new theoretical concepts are just starting to emerge. (…) The trust in the generative self-sufficiency of codes (in algorithmic and parametric processes and in the collection of big data) largely proceeded without any detours through the questions of formal meaning that played such a central role in the preceding decades." Emmanuel Petit, "Involution, Ambience, and Architecture," in *Log #29 "In Pursuit of Architecture"* (New York: Anyone Corporation Press, Fall 2013), p. 27.
18. See Humberto R. Maturana & Francisco J. Varela, *Autopoiesis and Cognition: the Realization of the Living* (Dordrecht: D. Reidel Pub. Co., 1980).
19. Patrik Schumacher, *The Autopoiesis of Architecture: A New Framework for Architecture* (Chichester: Wiley, 2011), p. 1.
20. See Geoffrey Broadbent & Anthony Ward, "Design Methods in Architecture," in *Architectural Association Paper* (New York: G. Wittenborn, 1969); and Alise Upitis, "Nature Normative: The Design Methods Movement, 1944–1967" (PhD diss., Cambridge, MA: MIT, 2008).
21. See, for instance, Christopher Alexander, *Notes on the Synthesis of Form* (Cambridge: Harvard University Press, 1964); George Stiny & James Gips, *Algorithmic Aesthetics: Computer Models for Criticism and Design in the Arts* (Berkeley: University of California Press, 1978); and John Frazer, *An Evolutionary Architecture* (London: Architectural Association, 1995).
22. See, for instance, Norbert Wiener, *Cybernetics; or, Control and Communication in the Animal and the Machine* (New York: MIT Press, 1961); Claude Elwood Shannon & Warren Weaver, *The Mathematical Theory of Communication* (Urbana: University of Illinois Press, 1949); and Marshal McLuhan, *Understanding Media: the Extensions of Man* (New York: McGraw-Hill, 1964).
23. See Hannah Arendt, *The Human Condition* (Chicago: University of Chicago Press, 1998); *Between Past and Future; Eight Exercises in Political Thought* (New York: Viking Press, 1968); *The Origins of Totalitarianism* (New York: Harcourt, Brace & World, 1966); and *The Life of the Mind* (New York: Harcourt Brace Jovanovich, 1978).
24. For example, see Ernesto Laclau & Chantal Mouffe, *Hegemony and Socialist Strategy: Towards a Radical Democratic Politics* (London: Verso, 1985); and Claude Lefort, "The Question of Democracy," in *Democracy and Political Theory* (Minneapolis: University of Minnesota Press, 1988). Refer to the bibliography for further titles.
25. See Jacques Rancière, "Democracy, Republic, Representation," *Constellations*, Vol. 13, No. 3 (2006); Jacques Rancière, *Disagreement: Politics and Philosophy* (Minneapolis: University of Minnesota Press, 1999); and Jacques Rancière, *The Politics of Aesthetics*, trans. Gabriel Rockhill (London: Bloomsbury Academic, reprint edition 2013).

26. For example, see Bruno Latour, *We Have Never Been Modern* (Cambridge, MA: Harvard University Press, 1993); and Bruno Latour, *Reassembling the Social: An Introduction to Actor-Network-Theory* (Clarendon Lectures in Management Studies) (Oxford: Oxford University Press, 2007).
27. For Foucault's major insights in this area, see Michel Foucault, *Society Must Be Defended: Lectures at the Collège De France, 1975–76* (New York: Picador, 2003); Michel Foucault, *Security, Territory, Population: Lectures at the Collège De France, 1977–78* (New York: Picador, 2007); and Michel Foucault, *The Birth of Biopolitics: Lectures at the Collège De France, 1978–79* (Basingstoke: Palgrave Macmillan, 2008).
28. Foucault, *The Birth of Biopolitics*, p. 61.
29. Ibid., p. 121 [our emphasis].
30. François Lyotard, trans. Geoff Bennington & Brian Massumi, *The Post Modern Condition: A Report on Knowledge* (Manchester: Manchester University Press, 1984). Originally published in France in 1979 by Les Editions de Minuit.
31. Maurizio Lazzarato, *Signs and Machines: Capitalism and the Production of Subjectivity* (Los Angeles, CA: Semiotext(e), 2014); Michael Hardt & Antonio Negri, *Commonwealth* (Cambridge, MA: Belknap Press of Harvard University Press, 2009); Paolo Virno, *A Grammar of the Multitude: For an Analysis of Contemporary Forms of Life* (Cambridge, MA: Semiotext (e), 2003); Christian Marazzi, *The Violence of Financial Capitalism* (Los Angeles, CA: Semiotext(e), 2010); and Franco "Bifo" Berardi, *The Soul at Work: From Alienation to Autonomy* (Los Angeles, CA: Semiotext(e)/Foreign Agents, 2009).
32. Indispensible in this very recent discourse is the anthology edited by Armen Avanessian & Robin Mackay, *#Accelerate: The Accelerationist Reader* (Falmouth: Urbanomic, 2014).
33. Arjun Appadurai, *The Future as Cultural Fact: Essays on the Global Condition* (London & New York: Verso Books, 2013).

Chapter 2

The historical pertinence of parametricism and the prospect of a free market urban order

Patrik Schumacher

Introduction

To respond to manifest societal trends, i.e., technological, socioeconomic, and political trends, is a vital capacity of architecture. However, this response must be an architectural rather than a political response. Debating politics within architecture can only concern the identification of manifest political trends. It can never be political debate, i.e., never a participation in the political controversies themselves. Architecture cannot substitute itself for the political process proper and must leave politico-economic innovations and the elaboration of radical politico-economic alternatives to the political and economic arenas. Architecture has no capacity to resolve political controversy. Political controversy and activism would overburden and explode the discipline. However, architecture can and must respond to transformative historical developments that become manifest within the economy and the political system. Architecture can only react with sufficient unanimity and collective vitality to dominant political agendas that already have the real power of a tangible political force behind them. Architectural discourse must develop innovative architectural responses to these historical transformative trends. This task raises the question of the historical adaptive pertinence of the various competing architectural tendencies, and in particular the question of parametricism's historical pertinence might be posed.

The political and socioeconomic premises of parametricism are found in the advancing processes of post-Fordist restructuring, globalization, market liberalization, and democratization. The methods of parametricism operate in line with the demands of post-Fordist flexible specialization and deliver attendant economies of scope. What is less clear is how the enhanced spatio-morphological ordering ambitions of parametricism can be compatible with the tendency that urban development is increasingly driven by market forces in the context of an eroding public planning capacity. The resultant urban disarticulation leads to the much bemoaned lack of identity and sense of urban chaos of the global "garbage spill" urbanization.

The increasing societal diversity paradoxically leads to white noise sameness rather than divers urban identities.[1] The imposition of visual order through top-down planning is no longer economically viable and in any event would blunt rather than reveal the underlying social complexity. In the face of the problem of disorienting urban disarticulation, parametric urbanism posits its vision of the bottom-up emergence of a complex variegated urban order whereby the different geographical, climatic, industrial, and cultural specificities of development sites become the starting point for the self-amplifying, path-dependent emergence of legible urban identities. Below, the architectural, disciplinary preconditions of such a scenario will be elaborated.

The theoretical starting point of parametricism is thus the acceptance of the pervasive historical socioeconomic trends of post-Fordist restructuring, globalization, and market liberalization—inclusive of market-led urban land-use allocation—and the eagerness to engage the challenges that these trends pose for architecture and urbanism's adaptive innovation, challenges that architecture and urbanism so far have failed to meet.

Locating architecture's specific criticality: the false pretense of "political architecture"

The stance of parametricism is sharply critical of current architectural and urban design outcomes. However, parametricism's stance is implicitly affirmative with respect to the general societal (social, economic, and political) trends that underlie the criticized urban outcomes. This implicit affirmation is a necessary condition of professional engagement with social reality.[2] The currently fashionable concept of a "critical" or "political" architecture as a supposed form of political activism must be repudiated as an implausible phantom.

The notion of a political architecture has transformed from a tautology to an oxymoron. In pre-Modern times, fortresses, palaces, and other major monuments were constituents of the political system, as were religion, the law, and the economy. In Modern times, architecture and politics have become separate autopoietic function systems. This raises the question of their proper relationship, their mutual observation, engagement, and adaptation.

The paradigmatic examples from the early 1920s and the late 1960s that give meaning to the notion of politically engaged architecture were born in the exceptional condition of social revolution (or impending social revolution). During such periods, everything was being politicized: the law, the economy, education, architecture, and even science. The autonomy of the functional subsystems of society was temporarily being suspended. During normal times, architecture and politics are separated as autonomous discursive domains. They are autopoietic function systems within a world civilization that is now primarily ordered via functional differentiation. Representative democracy is the form the political system tends to take in the most advanced states within functionally differentiated world society. Representative democracy professionalizes politics and regularizes the channels of political influence, negotiation, and collectively binding decision-making. The specialized, well-adapted channels of political communication absorb and bind all political concerns. Art, science, education, architecture, etc., are released from the burden of becoming vehicles of political agitation. The more this system consolidates, the longer this division of labor within society works, the more false and out of place rings the pretense of "political architecture." The term has become an oxymoron—at least until the emergence of the next revolutionary situation, when a new socio-political upheaval re-politicizes all aspects and arrangements of society. At *that* stage—within the throes of a genuine social revolution—we can expect the (temporary) meltdown of these distinctions and their underlying differentiation. Then we are no longer concerned with politics in the operationally defined sense that this term currently denotes.

Until then it is certainly *not* architecture's societal function to initiate or promote political agendas. Those who feel that a radical political transformation of society is a prior condition of any meaningful architectural project, and therefore want to debate and resolve political questions, must do so within the political system. Only there can they really form sustained political convictions and test the power of their arguments.

The concept of a "critical" or "political" architecture is either due to the delusion that the revolution has arrived or an atavism that dreams about small "brotherhood"-style societies without the functional division of societal realms that characterizes contemporary world society. This desire to collapse systemic distinctions that order society—like the desire for the atavistic fusion of architecture, economy, and politics—leads toward totalitarianism.[3]

The distinctions between architecture, art, science, economy, and politics are reflecting the historically specific, current ordering of societal discourses/practices. The failure to understand and reckon with these distinctions leads to self-defeating projects. Although philosophers or self-styled activists might wish to abolish or re-mix these categories, society reinforces these distinctions daily. While philosophy hovers above these domains unboundedly, real activists and professionals are bound by them, and for good reasons: society can this way parallel-process a much more complex and accelerated evolution via the co-evolution of autopoietic, functional subsystems that stimulate each other's evolution without being subject to a single master discourse. The evolved complexity of world society and the world division of labor cannot tolerate de-differentiation. De-differentiation would imply a crippling loss of productivity. Holisms break at this complexity barrier, as the catastrophic experiences of twentieth-century totalitarianism demonstrate. Actors who want to act effectively need to know whether their project is an art project, a sociology project, a political project, or a design project, etc. This is a condition of effective action in twenty-first-century world society.

To be effective, the innovative architect does not require any explicit political position. Indeed, the stating of political preferences and affiliations would be inappropriate within a professional context, just as we would not want to be bothered by the political position of our doctor or lawyer. This professional attitude of restricting one's communication to the domain of relevancy of one's competency is a necessary requirement of competent communication today. To violate this social norm amounts to the abdication of one's social role as a professional and implies the termination of the intended communicative nexus. It is clients that take responsibility for the (potentially controversial) political status of projects. The political meaning of the project is attributed to the client. Although a project might be controversial, the fact that it can go ahead at all implies that it is in principle consistent with the prevailing, legitimate constitution of society. In this sense, it is by default within the bounds of "mainstream politics."

Although the political muteness and implied mainstream position posited here is pervasive in practice, it is not adequately reflected in the discipline's explicit theoretical reflections. Here, the idea predominates that the architect should be "politically critical" and should not just acquiesce to the client's agenda. It is often said that the architect should assume a role and societal responsibility that reaches beyond the client's interest to take into account all stakeholders and to regard the public interest at large. The client's merely "commercial interests" are deemed problematic (despite the fact that in an advanced, money-mediated society all life interests take the form of "commercial" interests, and this includes the architect's own livelihood).[4]

Within functionally differentiated societies, the architect is only answerable to his client, and it is the client who is obliged (by political imposition) to take care that all stakeholders' interests considered politically relevant are recognized, and who instructs the architect accordingly. That all legitimate interests are taken into account is the client's, the planner's, and ultimately the court's responsibility. Within the given legal and political constraints, the market regulates the programmatic allocation of land resources to the effectively demanded social uses as anticipated by entrepreneurs. Architects interpret these contents spatially and formally—via spatial organization and formal articulation—to allow the flourishing of those specific social life processes that the client or hosting institution would like to host, and to simultaneously safeguard the interests of all those stakeholders the client has instructed him to consider. Any further self-appointment of the architect as "guardian of the public interest" would be delusional, arbitrary, and simply unprofessional.

Parametricism: candidate epochal style for the twenty-first century

It is important to explode the delusional pretense of a "critical architecture" in order to clear the view and path toward viable architectural ambitions that can make a real contribution to the progress of world civilization, in line with the discipline's specific competency. If we architects gain a realistic grasp of architecture's real domain of competency, and thus its specific criticality, we can become indeed very ambitious. Here is a realistic ambition that is likely to be crowned by pervasive success: the total makeover of the physiognomy of the global built environment and the world of artifacts according to the principles of parametricism. Is this aiming too high? Not at all. Think back to the Bauhaus of the mid-1920s. Did not what they developed then deliver a total makeover of the physiognomy of the global built environment and the world of artifacts in the twentieth century? There is nothing that escaped the Bauhaus' thrust in the twentieth century. Today's reach of the globally integrated design discourse is much more pervasive, and its trends disseminate more rapidly. More than ever, the totality of world production has to pass through this needle's eye of a discursively determined design paradigm. Everything that is made is being designed by a designer, and thus by a participant of our global design discourse.

Of course, this total spatio-morphological makeover is only a part of total society. This makeover has to co-evolve with the parallel processes of change in world economy, world science, technology, world politics, etc. The ambitions of parametricism update the ambitions of the modern movement.[5]

The key historical category that motivates and calls for parametricism's takeover from Modernism is "post-Fordist network society" as distinct from the prior era of Fordist mass society. In *The Autopoiesis of Architecture*, the author has elaborated a theory of styles within which the concept of epochal styles implies a historical alignment with societal (socioeconomic) epochs. Architecture emerged from tradition-bound building as a differentiated, consciously innovative, theory-led discipline in the Renaissance, and advanced—in co-evolution with the other societal subsystems like science, the economy, politics, etc., that started to be differentiated at the same time—via the progression of epochal styles. (Postmodernism and Deconstructivism are transitional rather than epochal styles, transitional episodes between Modernism and Parametricism, like Art Nouveau and Expressionism were transitional styles on the way to Modernism.) The epochal location of parametricism can be succinctly characterized by the following figure:

Building / Architecture	Society / Socio-Economic Epoch
Tradition-Bound Building	
Medieval Vernacular	Feudalism
Romanesque	Feudalism
Transition	
Gothic	Feudalism + Rising Cities
Architectureal History: Epochal Styles	
Renaissance	Early Capitalism / City States
Baroque	Mercantilism / Absolutism
Neo-Classicism / Historicism	Bourgeois Capitalism / Nation States
Modernism	Fordism / International Socialism
Parametricism	Global / Post-Fordist Network Society

Figure 2.1 Epochal alignments of styles.

Parametricism is valid and vital whether post-Fordist socioeconomic restructuring proceeds within a social democratic or a liberal (or even libertarian) political frame, just as Modernism was compatible with both capitalism and socialism in the twentieth century.[6] On both sides of The Iron Curtain, modernist architecture had to adapt to the respective political and economic conditions. The essential characteristics of Modernism survived these divergent adaptations.[7] In the same way, parametricism's historical pertinence will have to prove its robustness with respect to the particular and potentially divergent adaptations that global post-Fordism will engender in the political realm in the different parts of world society.

Figure 2.2 Medieval town.

Figure 2.3 Palmanova, Renaissance.

Figure 2.4 The Palace of Versailles, the Grand Trianon, ca 1668. Artist: Patel, Pierre (1605–1676) (Photo by Fine Art Images/Heritage Images/Getty Images).

Parametricism is the contemporary style that is most vigorously advancing its design agenda on the basis of computationally augmented, parametric design techniques. It is a widespread paradigm and global movement within contemporary architecture that emerged and gathered momentum during the last fifteen years. The author is an active participant in the advancement of this movement via teaching arenas like the AA Design Research Lab and via the designs and buildings of Zaha Hadid Architects. The movement—the only truly innovative direction within contemporary architecture—has by now sufficiently demonstrated its capacity to credibly aspire to become the universally recognized "best practice" approach to architectural and urban design globally. Parametricism is ready to make an impact: to transform the physiognomy of the global built environment and the totality of the world of designed artifacts, just like Modernism did in the twentieth century.

As a conceptual definition of parametricism, one might offer the following formula: Parametricism implies that all architectural elements and compositions are parametrically malleable. This implies a fundamental ontological shift within the basic, constituent elements of architecture. Instead of the classical and modern reliance on ideal geometrical figures—straight lines, rectangles,

Historical pertinence of parametricism

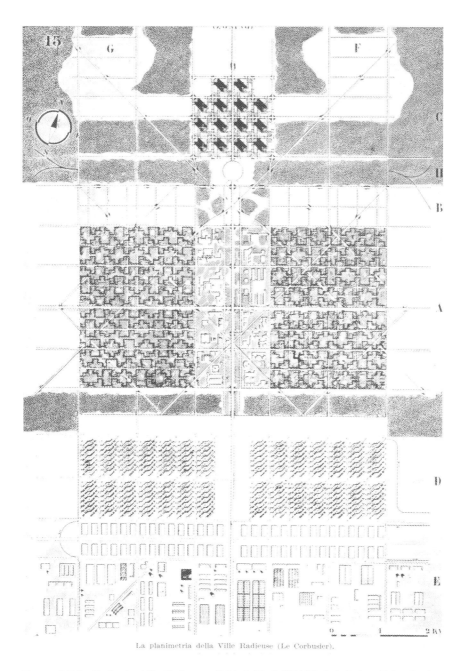

La planimetria della Ville Radieuse (Le Corbusier).

Figure 2.5 Le Corbusier, Ville Radieuse, 1924. © F.L.C./ADAGP, Paris/Artists Rights Society (ARS), New York 2014.

Figure 2.6 Zaha Hadid Architects, Istanbul Master Plan, 2007.

as well as cubes, cylinders, pyramids, and (semi-)spheres—the new primitives of parametricism are animate (dynamic, adaptive, interactive) geometrical entities—splines, nurbs, subdivs, particle-spring systems, agent-based systems, etc.—as fundamental "geometrical" building blocks for dynamical compositions that can be made to resonate with each other (and with contextual conditions) via scripts.

In principle, every property of every element or complex is subject to parametric variation. The key technique for handling this variability is the scripting of functions that establish associations between the properties of the various elements. However, although the new style is to a large extent dependent upon these new design techniques, the style cannot be reduced to the mere introduction of new tools and techniques. What characterizes the new style are new ambitions and new values—both in terms of form and in terms of function—that are to be pursued with the aid of the new tools and techniques. Parametricism pursues the very general aim to organize and articulate the increasing diversity and complexity of social institutions and life processes within post-Fordist network society. For this task, parametricism aims to establish a complex variegated spatial order. It uses scripting to lawfully differentiate and correlate all elements and subsystems of a design. The goal is to *intensify the internal interdependencies* within an architectural design, as well as the *external affiliations and continuities* within complex, urban contexts. Parametricism offers a new, complex order via the principles of *differentiation and correlation*.

This general verbal and motivational definition of parametricism can and must be complemented by an operational definition. It is necessary to operationalize the intuitive values of a style in order to make its hypotheses testable, to make its dissemination systematic so that it can be exposed to constructive criticism, including self-critique.

The operational definition of a style must formulate general instructions that guide the creative process in line with the general ambitions and expected qualities of the style. A style is not only concerned with the elaboration and evaluation of architectural form. Each style poses a specific way of understanding and handling functions. Accordingly, the operational definition of parametricism comprises both a formal heuristics—establishing rules and principles that guide the elaboration and evaluation of the design's formal development and resolution—and a functional heuristics—establishing rules and principles that guide the elaboration and evaluation of the design's social functionality.

For each of these two dimensions, the operational definition formulates the heuristics of the design process in terms of operational taboos and dogmas specifying what to avoid and what to pursue. At the same time, these heuristic design guidelines provide criteria of self-critique and continuous design enhancement.

Operational definition of parametricism:
Formal heuristics:
Negative principles (taboos): avoid rigid forms (lack of malleability)
avoid simple repetition (lack of variety)
avoid collage of isolated, unrelated elements (lack of order)
Positive principles (dogmas): all forms must be variable and adaptive (deformation = information)
all systems must be differentiated (gradients)
all systems must be interdependent (correlations)
Functional heuristics:
Negative principles (taboos): avoid rigid functional stereotypes
avoid segregative functional zoning
Positive principles (dogmas): all functions are parametric activity/event scenarios
all activities/events communicate with each other

The avoidance of the taboos and the adherence to the dogmas deliver complex, variegated order for complex social institutions. These principles outline pathways for the continuous critique and improvement of the design. The designer can always increase the coherence and intricacy of his/her design by inventing further variables and degrees of freedom for the composition's primitive components. There is always scope for the further differentiation of the arrays or subsystems that are made up by the elemental primitives. This differentiation can be increased with respect to the number of variables at play, with respect to the range of differences it encompasses and with respect to the fineness and differential rhythm of its gradients. There is always further scope for the correlation of the various subsystems at play in the multisystem setup. Ultimately, every subsystem will be in a relation of mutual dependency with every other subsystem, directly or indirectly. The number of aspects or properties of each subsystem that are involved in the network of correlation might be increased with each design step. Further, there is always the possibility (and often the necessity) to add further subsystems or layers to the (ever more complex and intricate) composition. Also, it is always possible to identify further aspects or features of the (principally unlimited) urban context that might become an occasion for the design to register and respond to. Thus, the context sensitivity of the design can be increased with every design step. Thus, the heuristics of parametricism direct a trajectory of design intensification that is, in principle, an infinite task and trajectory. There is always a further possibility pushing up the intensity, coherence, and intricacy (and beauty) of the design. As the network of relations tightens, each further step becomes more elaborate, more involved as all the prior subsystems and their trajectories of differentiation should be taken into account. Arbitrary additions show up conspicuously as alien disruption of the intricate order elaborated so far. Each additional element or subsystem that enters the composition at a late, highly evolved stage challenges the ingenuity of the designer, and more so the more the design advances. The complex, highly evolved design assumes more and more the awe-inspiring air of a quasi-natural necessity. However, the design remains open-ended. There can be no closure. The classical concepts of completeness and perfection do not apply to parametricism. Parametricism's complex variegated order does not rely on the completion of a figure. It remains an inherently open composition and design trajectory.

In the perspective of architecture, and specifically in the perspective of contemporary parametric design, contemporary society is a vast panoply of parametrically variable event scenarios. This formula spells the program dimension of the built environment in terms of the parametricist paradigm. So far, parametricism has primarily focused on formal correlation; the correlation of formal–spatial subsystems. However, it is pertinent to expand parametricism's key concept of correlation to include form–function relations; i.e., including the correlation of the patterned built environment with the patterns of social communications that unfold within it. This is meaningful because the same computational techniques that operationalize the concept of formal correlation can now be applied to form–function correlations. How is this possible?

The functional heuristics of parametricism, as defined above, propose to conceive of the functions of spaces in terms of dynamic patterns of social communications; i.e., as parametrically variable, dynamic event scenarios, rather than static schedules of accommodation that list functional stereotypes. It has now become possible to model the functional layer of the built environment and thus incorporate it into the design process. This is made possible by computational crowd-modeling techniques via agent-based models. Such models reproduce or predict collective patterns of movement from the aggregation of individual agents presumed to navigate their environment according to defined rules. Tools like MiArmy or AI.implant (available as plug-ins for Maya), or

Massive now make behavioral modeling within designed environments accessible to architects. Agent modeling should not be limited to crowd circulation flows, but should encompass all patterns of occupation and social interaction in space. The agents' behaviors might be scripted so as to be correlated with the configurational and morphological features of the designed environment; i.e., programmed agents respond to environmental clues. Such clues or triggers might also include furniture configurations as well as other artifacts. The idea is to build up dynamic action–artifact networks. Colors, textures, and stylistic features, which together with ambient parameters (lighting conditions), constitute and characterize a certain territory that might influence the behavioral mode (mood) of the agent. Since the "meaning" of an architectural space is the (nuanced) type of event or social interaction to be expected within its territory, the new tools allow for the refoundation of architectural semiology as parametric semiology. This implies that the meaning of the architectural language can enter the design medium (digital model). The semiological project implies that the design project systematizes all form–function correlations into a coherent system of signification. A system of signification is a system of mappings (correlations) that map distinctions or manifolds defined within the domain of the signified (here the domain of patterns of social interaction) onto distinctions or manifolds defined within the domain of the signifier (here the domain of spatial positions and morphological features defining and characterizing a given territory) and *vice versa*. The system of signification works if the programmed social agents consistently respond to the relevantly coded positional and morphological clues so that expected behaviors can be read off the articulated environmental configuration.[8]

A unified style argued for within a unified theory of architecture

The author (who originally coined this term in 2007–2008) argues that parametricism is the only truly innovative direction within architecture and should be promoted as the only credible candidate aspiring to become the unified epochal style for architecture, urbanism and all the design disciplines for the twenty-first century. This thesis is being argued for within a comprehensive, unified theory of architecture/design (the theory of architectural autopoiesis) that is embedded within an overarching theory of society (Niklas Luhmann's social systems theory).[9]

Clarity and unity of agenda are necessary preconditions for effective collective action.[10] Such clarity and unity must not be based on prejudice or a simplistic worldview. The world within which architecture must try to redefine its role and agenda seems much more complex and uncertain than the world in which architectural modernism made its mark fifty years ago. Many have come to believe that this world is too fragmented and contested to allow for a unified collective architectural agenda analogous to the modern movement. This skepticism within architecture coincides with the general distrust for grand comprehensive theories in philosophy and the social science since the crisis of Marxism as well as of development and modernization theories in general. The 1970s exposed the naivety of those theories' expectations of global industrialization, democratization, the expansion of general social welfare, and the smoothing of business cycles via democratic regulation and central steering. The world had become a rather more complex and unpredictable place. This new condition was reflected in the new theoretical sophistication of poststructuralist philosophy. New loops of theoretical reflection—the reflection upon language, discourse, audience, institutional interests, etc.—circumscribed and relativized the substantial claims that might still be made in social and cultural analysis. However, the ability to navigate the new cultural complexities was accompanied by a relativist stance that was skeptical toward attempts at overarching theoretical synthesis. So, there is no comprehensive poststructuralist theory of contemporary society. Instead, we witness a proliferation of perspectives and areas of study. At the same time, Modernism in architecture/design gave way to a plurality of styles and approaches.

However, the increased complexity and diversification of social phenomena, more than ever, unfolds within a single integrated social world. Thus there remains—more than ever—the possibility (and necessity) of a unified theory; a theory that is much more complex than earlier theories, and that includes the new poststructuralist insights and loops of reflection.

There is indeed such a social theory, a theory that can cope with this new level of complexity and uncertainty. This is Niklas Luhmann's theory of "functionally differentiated society," embedded in his "social systems theory," based on complexity theory and the theory of autopoiesis. My unified theory of architecture explicitly builds on Luhmann's theory and can be read as a new component within his theoretical system.

Many have come to believe that the pluralism of styles and perspectives that emerged in the 1970s and 1980s is an inherent and inevitable characteristic of our epoch, and that a globally shared architectural agenda analogous to the modern movement is no longer possible. Against this stands the fact that a global convergence of design research efforts has gathered sufficient momentum within the architectural avant-garde over the last fifteen years to make the emergence of a new unified paradigm and agenda in analogy to Modernism plausible today. In 2008 I proposed a name for this movement—parametricism—and started my attempts to summarize its novel features, methodologies, and values. As a committed participant, I also tried to explicate its rationality, advantages, and preliminary achievements in the light of the current "historical" condition: the globalized, knowledge and network society. Due to the 2008 financial crisis and its economic aftermath—the Great Recession—the proliferation of parametricism has been much slower than one might have expected six years ago. However, further progress has been made in the movement's evolution from an (ongoing) avant-garde design research agenda to a movement with the strategic agenda of global implementation across all scales and program categories. At least this is the author's ambition. Its viability is demonstrated by the dramatic expansion of Zaha Hadid Architects in scale, scope, and global reach.[11]

Is this ambition and claim toward the global implementation of the new paradigm not contradicted by the diversity of climatic, socioeconomic, and cultural conditions? My answer to this often-posed question is that differentiation and local adaptation is the very essence of parametricism. The abstractness and thus open-endedness of its general heuristic principles guarantee the adaptive versatility of its solution space. While the world is more diverse and differentiated—across countries and continents, as well as within its megacities—it also is more interconnected and integrated than ever, so that talk of a single world society becomes ever more justified. Thus no region, culture, or subculture can remain secluded from the most advanced, global best-practice architectural paradigm. This is already borne out by the fact that in nearly all countries of the world both projects and protagonists of parametricism can be found.

As hinted at above, parametricism is the only credible candidate for a new global epochal style for architecture, urbanism, and the design disciplines since Modernism's crisis and demise more than thirty years ago. Its emergence so far is already a significant fact of architectural history. However, an avant-garde style is an emergent, evolving discursive phenomenon in which many contributors and voices coalesce. Its identification, naming, and theoretical explication is just one more discursive event within its evolution. This was my contribution via my writings since 2008 and via Volume 1 of my treatise *The Autopoiesis of Architecture*. Volume 1 is subtitled *A New Framework for Architecture*.[12] It charts, rationalizes, and historicizes what has been achieved. By itself it does not yet formulate a decisive, future-oriented agenda for architecture's further progress. The formulation of an agenda for architecture's further advance and the attempt to upgrade the discipline's intellectual and methodological resources in order to meet the challenges posed and to exploit the opportunities afforded by contemporary civilization, this is the ambition of the second volume of my treatise. Its subtitle reads *A New Agenda for Architecture*.[13]

The key categories in Luhmann's theoretical edifice are the concept of "communication" and the concept of "social system." Social systems are systems of communications. This makes sense in a world where all problems are now first of all problems of communication. World society is understood as the totality of all communications. These communications are self-organized into autopoietic[14] systems primarily according to the principle of functional differentiation. Accordingly, my theory of architecture theorizes the discipline of architecture as an evolving system of communications (discourse) that takes universal and exclusive responsibility for the innovation of the built environment, in functional differentiation to engineering, science, art, politics, law, economy, etc. The specific responsibility of architecture/design concerns the social/communicative functionality (in distinction to the technical/engineering functionality) of the built environment. Spaces are theorized as framing communications that function as invitations and premises for all the interactions that (are meant to) take place within them. Each territory/frame is embedded within a system of frames that can be understood as a system of signification.

Society can only evolve with the simultaneous ordering of space. The life process of society is a communication process that is structured by an ever more complex and richly diversified matrix of institutions and communicative situations. The built environment orders and stabilizes this matrix of institutions and makes it legible. The innovation of the built environment participates in the expansion, differentiation, and integration of this network of communicative situations. The built environment with its complex matrix of territorial distinctions is a giant, navigable, information-rich interface of communication.

Within contemporary network society (information society, knowledge economy), total social productivity increases with the density of communication. Contemporary network society demands that we continuously browse and scan as much of the social world as possible to remain continuously connected and informed. We cannot afford to withdraw and beaver away in isolation when innovation accelerates all around. We must continuously recalibrate what we are doing in line with what everybody else is doing. We must remain networked all the time to continuously ascertain the relevancy of our own efforts. Rapid and effective face-to-face communication remains a crucial component of our daily productivity. Telecommunication cannot replace face-to-face and group communication or the browsing of a dense urban environment. The importance of the built environment further increases as *mobile* telecommunication unchains us from our desks and releases us into the space of the city. The whole built environment becomes an interface of multimodal communication, and the ability to navigate dense and complex urban environments is an important aspect of our overall productivity today. Our increasing ability to scan an ever-increasing simultaneity of events, and to move through a rapid succession of communicative encounters, constitutes the essential, contemporary form of cultural advancement. The further advancement of this vital capacity requires a new built environment with an unprecedented level of complexity, a complexity that is organized and articulated into a complex variegated order of the kind we admire in natural self-organized systems. The city is a complex densely layered text and a permanent broadcast. Our ambition as architects and urban designers must be to spatially unfold more simultaneous choices of communicative situations in dense, perceptually palpable, and legible arrangements. The visual field must be rich in interaction opportunities and information about what lies behind the immediate field of vision.

My thesis is that the built environment should be conceived and designed as a three-dimensional, 360-degree, layered interface of communication. It can communicate more as more becomes simultaneously visible. But that is not enough. Its communicative capacity depends on the coherency of its internal order so that what is visible allows for inferences about what is invisible, or not yet visible. This depends on the consistency of its form–function correlations, so that a positional or morphological distinction or difference makes a predictable difference in terms of expected social interaction patterns or social functions. Thus, the built environment's communicative capacity is enhanced the more the employed architectural order and morphology is designed as a coherent system of signification.

The prospect of a free market urban order

All top-down bureaucratic attempts to order the built environment are bankrupt. The experiences of the last thirty-five years indicate that within the "postmodern condition" (post-Fordist network society) all political attempts to intervene in spontaneous urban development processes lead to wasteful distortions, delays, underutilization, shortages, and inflated real estate prices. Only market processes can process the new diversity and complexity of information, and generate the knowledge required to deliver land and real estate resources reliably to productive and desired uses, avoiding wasteful misallocations. However, while laissez-faire development can deliver a socially (market) validated program mix and program distribution, it seems bound to produce visual chaos in the urban dimension. This visual disorder is not only ugly and distracting; it is disorienting, and thus compromises the social functionality of the built environment.

The articulation of a legible spatial order—the architect's core competency—is itself a vital aspect not only of the city's "liveability" but also of its economic productivity.[15]

The question is thus posed: How can the vital desire for urban order, legibility, and identity be reconciled with a free market that includes an equally unrestrained artistic freedom? In short: How can entrepreneurial and artistic freedom lead to urban order? My answer is twofold:

1   Freedom and order can coincide in private planning.
2   Freedom and order beyond the bounds of private planning can emerge via the discursive convergence of the design disciplines toward a new epochal style: parametricism.

Around 1980, large-scale state-planned and subsidized urban expansions vanished in the advanced economies, and with them vanished the disciplines of urbanism and physical urban planning. Planning was henceforth confined to operate negatively by means of restricting private actors. The disappearance of urbanism is a moment in the crisis of Modernism, itself but a moment in the crisis of Fordism understood as the era of the mechanical mass production of a national consumption standard within a planned or mixed economy, under the auspices of either socialist or social-democratic welfare politics. The last thirty years since have been marked by the reassertion of market forces and globalization under the auspices of neoliberal politics. Since 1980 we live in the era of a market-led post-Fordist socioeconomic restructuring. The readmission of international market forces and entrepreneurship—a reaction to the grinding 1970s socioeconomic stagnation of economic interventionism and welfare-ism—combusted with the versatile productive potentials of the microelectronic revolution to unleash a new socioeconomic dynamic: the emergence of post-Fordist network society. Lifestyle diversification and the new diversity in products and services made economically viable by the new design and production systems engaged in mutual amplification. The diversity of new enterprises, coupled with accelerating cycles of innovation (made viable by the new technologies and expanded markets), engendered a much differentiated and intensified societal communication. The planned decentralization via mute, monotonous, zoned satellite settlements separating sleeping silos from industrial estates was no longer a viable recipe for societal advancement. In terms of urban development, this implied the return to the historic centers with individual incisions as well as a deregulated laissez-faire sprawling beyond the bounds of emerging megacities.[16] Both tendencies can be described as forms of collage, the antithesis of planned or designed development. The result is what I have called "garbage spill urbanization." This mode of development is certainly better adapted to the new socioeconomic dynamics than the bankrupt simplistic order of Modernist planning and urbanism. However, it produces a disorienting visual chaos that compromises the vital communicative capacity of the built environment. While the new diversity and open-endedness of post-Fordist social phenomena is being accommodated, the unregulated agglomeration of differences produced the global effect of "white noise" sameness everywhere without allowing for the emergence of distinct urban identities within a legible urban order.

This phenomenological disarticulation of the emergent organizational complexity hampers the full potential for complex social organization and communication. Social functionality depends as much on subjective visual accessibility as it depends on objective physical availability. Architects should recognize this instrumentality of visual appearances as a key moment of their core competency and task. Social cooperation requires that specifically relevant actors find each other and configure within specific communicative situations. This insight motivates architectural attempts to articulate a complex variegated urban order that allows for the intuitive navigation and orientation within an information-rich built environment that makes its rich offerings visually accessible. That is the design agenda of parametricism and parametric urbanism. There is no doubt that the new computational ordering devices like gradients, vector fields, and the methods of associative modeling and geometric data field transcoding allow designers to generate intricately ordered urban morphologies with distinct identities that could, in principle, make a much larger amount of programmatic information perceptually tractable. However, the question arises how this desired increase in urban order could be implemented in the face of a receding state planning apparatus?

One obvious way in which the vacuum left by state planning can be filled is by means of what might be called "private planning"; i.e., by a process whereby private development corporations or consortiums unify larger and larger development areas within a coherent market-controlled urban business strategy. This emerging phenomenon allows for the strategic, value-enhancing garnering and steering of programmatic synergies, while remaining continuously sensitive and flexible with respect to market and thus end-user validation; i.e., without suffering from the rigidities and blockages that hamper state efforts subject to all-too slow political and bureaucratic procedures, ideologies and rent-seeking special interest groups. "Planning" for a complex societal dynamic is inherently uncertain and thus speculative. The degree of flexibility that such a plan should have is one more uncertain speculation, a risky business that is being entrusted to entrepreneurs and investors with their own skin in the game, rather than to (essentially indifferent) bureaucrats. Private planning seems to have a better prospect to succeed in this fast-paced, dynamic environment without central control systems. Only market-based decisions, guided by up-to-date price signals, have a chance here to process the best relevant information and guarantee due diligence. A system of private planning—operating under the market's selection pressure of profit and loss—can process the information condensed into price signals in order to achieve a rational, economizing allocation of urban resources within a coherent, synergetic mix of most valued uses. It does so by aligning all vital actors' incentives with a relative urban value maximization, and thus most productive and desired end-user utilization. The degree of territorial comprehensiveness of such synergy-hunting private planning efforts is an empirical question that can only be answered by the unleashed market processes themselves. So far, we can observe two relevant empirical facts with respect to this question. First, it must be observed that even the most atomized market-based development process—where "planning" is confined to individual isolated parcels—has a far higher degree of adaptive pertinence with respect to post-Fordist network society and is generating more economic and urban vitality than the state-planned urban expansions of yesteryear. The second fact is that—while isolated insertions continue—we can observe a tendency toward attempts to merge and integrate parcels within the historical centers and toward larger and larger privately master-planned development sites within the wider expanse of the global megacities where development is concentrated. In this sense, private planning is on the rise and thus affords opportunities for visual as much as programmatic integration.[17]

Is the degree of order that parametric urbanism aspires to possible beyond the level of integration achievable via private planning? More generally: is urbanism at all possible in the face of free market dynamism? Deconstructivism can be looked at as the aesthetic ideology of this urban process of "garbage spill" collage. Like the move from classical architecture to Modernism, the move from Modernism to Deconstructivism and collage delivered an expansion of degrees of freedom and versatility (to accommodate a more complex society) that was paid for by a relaxation or rejection of rules of composition; i.e., of means of ordering, and thus a resultant degeneration of the visual order. Parametricism is the first style that delivers further degrees of freedom and versatility in conjunction with a simultaneous increase in its ordering capacity via new compositional rules like affiliations, gradients, and associative logics. In principle, all design moves are now rule-based and thus have the potential to enhance the visual order, and thus the legibility, of the built environment in the face of an increased complexity.

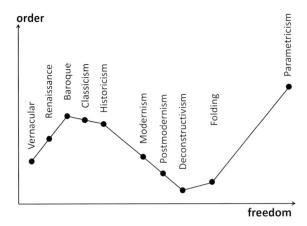

Figure 2.7 The simultaneous enhancement of freedom and order: inversion of architecture's entropy law.

If we look at the historical progression of styles, we find that the last 300 years established what we might call architecture's entropy law: all gains in terms of design freedom and versatility have been achieved at the expense of urban and architectural order; i.e., increases in versatility had to be bought by a progressive degeneration of architecture's ordering capacity. The increase of degrees of freedom established via the enrichment of architecture's formal-compositional repertoire was the paramount criterion of progress in architecture's pursuit of matching the requisite variety of societal complexity. Order was progressively eroded. This long trend of a negative correlation of freedom and order can be reversed under the auspices of parametricism. Parametricism offers the simultaneous increase in freedom and order and thus inaugurates a new phase of architectural neg-entropy. Parametricism's radical ontological and methodological innovation translates into a massive leap in both dimensions of architectural progress considered here; i.e., it entails an unprecedented expansion of architecture's compositional freedom and versatility and an unprecedented leap in architecture's ordering capacity through the deployment of algorithms and associative logics. This reversal of architecture's entropy law, this new ordering capacity or architectural neg-entropy, is the critical factor in architecture's potential to halt the ongoing urban disarticulation of the world's built environments. However, this factor can only come into play if parametricism achieves hegemony as the unified, epochal style of the twenty-first century.

The market process is an evolutionary process that operates via mutation (trial and error), selection, and reproduction. It is self-correcting and self-regulating, leading to a self-organized order. Thus we might presume that the land-use allocation and thus the programmatic dimension of the urban and architectural order are to be determined by architecture's private clients within a market process that allocates land resources to the most valued uses. However, in the absence of stylistic and methodological coherence we cannot expect the underlying programmatic order to become legible as a spatio-morphological order. For this to happen, we must presume a hegemonic stylistic and methodological paradigm that has the versatility and ordering capacity to translate the social order into a spatial order. A shared paradigm offers the prospect of coherence across multiple authors working for multiple clients. No controlling hand needs to be presupposed, no political enforcement. All that is required is the further discursive convergence towards a unified style that can deliver the combination of versatility (freedom) and order. The only viable candidate for the next hegemonic epochal style is thus parametricism.

Figure 2.8 Parametricism: complex variegated order via multiauthor coherence, Studio Hadid, Yale University, 2013.

Neither a hegemonic Postmodernism nor a hegemonic Deconstructivism could overcome the visual chaos that allows the proliferation of differences to collapse into global sameness ("white noise"). Both Postmodernism and Deconstructivism operate via collage; i.e., via the unconstrained agglomeration of differences. Only parametricism has the capacity to combine an increase in complexity with a simultaneous increase in order, via the principles of lawful differentiation and multisystem correlation. Only parametricism can overcome the visual chaos and "white noise" sameness that laissez-faire urbanization produces everywhere. Parametricism holds out the possibility of a free market urbanism that produces an emergent order and local identity in a bottom-up process; i.e., without relying on political or bureaucratic power. The values and methodological principles of parametricism are prone to produce path-dependent, self-amplifying local identities. Its ethos of contextual affiliation and ambition to establish or reinforce continuities allows for the development of unique urban identities on the basis of local contexts, topography, climate, etc. Parametricist order does not rely on the uniform repetition of patterns as Modernist urbanism does. In contrast to Baroque or Beaux Arts master plans, parametricist compositions are inherently open-ended (incomplete) compositions. Their order is relational rather than geometric. They establish order and orientation via the lawful differentiation of fields, via vectors of transformation, as well as via contextual affiliations and subsystem correlations. This neither requires the completion of a figure nor—in contrast to Modernist master plans—the uniform repetition of a pattern. There are always many (in principle, infinitely many) creative ways to transform, to affiliate, to correlate. Parametricism thus holds out the prospect of a free market urban order. The functional (programmatic) dimension of this new urban order is being delivered by client-entrepreneurs competing and collaborating within the institutional framework of the global market process: neoliberalism. The formal (spatio-morphological) dimension of this order can be delivered by architects competing and collaborating under the auspices of a shared, discursively institutionalized, global best-practice paradigm: parametricism.

Notes

1. This paradoxical inversion of richness into sameness is due to the limitations of perceptual cognitive processing.
2. This affirmation is usually implicit rather than explicit as the architect's political positioning would disrupt his architectural discourse.
3. The author is skeptical about totalizing programs of revolutionary action, while recognizing the need for totalizing theories of society as frameworks of orienting self-localization.
4. The distinction of social concerns and commercial interests is a dangerous commonplace. The unleashing of commercial forces is what has delivered us from the savage

destitution and bone-breaking drudgery of older times. This much even Karl Marx understood when he sang the praises of a globalized capitalist dynamism in his *Communist Manifesto*. Commerce is a dynamic social process in which everybody pursues his or her vital life interests.
5. They update the ambitions of the Modernism of a Le Corbusier, rather than the ambitions of bolshevism and Lenin—the last credible but failed totalizing philosopher king (who morphed into the murderous totalitarian dictator Stalin).
6. I have come to believe that post-Fordist network society needs to go further along the current path of globalization and most importantly liberalization, the unleashing of individualism and individual liberty as a precondition of bottom-up self-organizing, emergent systems of social cooperation. So, as a politically thinking citizen I am not afraid of political argument, but I am not willing to make my architectural position and agenda dependent upon my personal political position, and I do not believe that the fate of my architectural project depends on the realization of my political preference for a society based on individual liberty rather than political control. I believe that parametricism is congenial to this outlook; however, its validity as global "best practice" does not depend on this particular political premise.
7. This universal validity of Modernism on both sides of The Iron Curtain became manifest after the direct political control of design decisions was relinquished after Stalin's death in 1953.
8. This project of an agent-based parametric semiology is the author's most recent design research project. The first results are promising but as yet tentative.
9. My underlying epistemology is radical constructivism, which I understand to be a variant of pragmatism. The theory reflects itself as a designed theoretical construction that sets itself as contingent and understands itself as a provisional attempt to integrate a multiplicity of insights. It evaluates itself in terms of its comprehensiveness, fruitfulness, and relevance for contemporary architectural practice.
10. When we speak of collective action today, we no longer mean organized or centrally coordinated action. What is implied here is a spontaneous convergence within an open-ended network of communications.
11. Zaha Hadid started in 1980. After twenty years, in 2000, Zaha Hadid Architects had only completed three small buildings and was employing twenty people. Currently, Zaha Hadid Architects is employing 450 people, working on about eighty projects worldwide, across all program categories, including many large-scale projects over 100,000 m^2 and several above 300,000.
12. Patrik Schumacher, *The Autopoiesis of Architecture, Vol.1: A New Framework for Architecture* (London: John Wiley & Sons Ltd., 2010).
13 Patrik Schumacher, *The Autopoiesis of Architecture, Vol.2: A New Agenda for Architecture* (London: John Wiley & Sons Ltd., 2012).
14. The autopoiesis concept is fundamental to Luhmann's system. Together with the concept of parallel (rather than hierarchical) functional differentiation, it implies the impossibility of a societal control center. World society evolves via the coevolution of

self-referentially enclosed autopoietic social subsystems.
15. The failure to grasp this instrumentality of the built environment's appearance has for too long hampered architecture's proactive pursuit of formal articulation as a key competency of the discipline. The crucial work on formal/aesthetic problems, which in practice takes up the larger part of the architect's design work, is being denigrated or denied in the discipline's self-descriptions. Architecture is responsible for the built environment's social (rather than technical engineering) functionality. Social functionality of the built environment largely depends upon its communicative capacity, which in turn is a matter of visual communication through the built environment's appearance.
16. One of the most detrimental types of restriction that still hampers urban productivity is arbitrary restriction on the density of land use. Economic vitality and productivity gains concentrate in the major high-productivity communication hubs that are the world's megacities and conurbations of today. In Western Europe and North America, political planning impediments unduly constrain this tendency and thus restrict potential synergies and productivity gains. The major cities and high-productivity conurbations like London, New York, and Silicon Valley would grow much more dense were it not for the political impediments. Massive land and real estate price differentials between the central world cities and other cities and regions reveal both the productivity and desirability differentials, on the one hand, and the political restrictions of supply in the centers, on the other hand. Internal price differentials between land-use categories within cities—like the gaping price difference between residential and office properties in London—reveal how what is left of municipal planning is at odds with real requirements as expressed in market demands.
17. The example of London's great estates offers an encouraging historical precedent here, a precedent of private, market-based, long-term urban asset management and private planning establishing an urban order inclusive of a visual architectural order.

Chapter 3

On numbers, more or less
Reinhold Martin

If we take "architecture," as we normally do, to mean the profession and the academic discipline that was codified in the European nineteenth century, it is difficult not to conclude that there would be no architecture without a specific set of techniques for measuring, enumerating, and calculating the dimensions of things in space. Among the characteristics that differentiate those techniques from older ones, such as the geometrical systems of the early modern period (i.e., from the Renaissance to the early Enlightenment), is the widespread use of technical drawings as coordinating instruments. Whether it meant calculating the complex ratios of the Beaux Arts or the intricate patterns of the Gothic Revival, by the end of the nineteenth century, to produce architecture was, by and large, to compile detailed sets of scale drawings that were correlated mathematically with one another and with the eventual building, such that an entire edifice could be inferred from combinations of measured lines, and numbers.

Much historical work remains to be done on the subject, but it is fair to say that what we call modern architecture is, to a certain extent, an outcome of such practices. Beyond visual allusions to machinery, or the use of mass-produced materials, or even the distillation of "abstract" geometries, architectural design was bound to the multiform processes of industrialization by new ways of drawing, measuring, tabulating, and annotating that belonged to the bureaucratization of intellectual labor, on the one hand, and to new regimes for establishing objectivity, on the other. As Hyungmin Pai has shown, by the first decades of the twentieth century, European and American architectural discourse had exchanged, as its discursive base, the epic lines and washes of the large-scale portfolio for the succinct, diagrammatic calculations of the technical drawing.[1] To the extent that today we may be in the midst of a similar modulation, in which digital computation slowly and unevenly gives new meaning to what the German sociologist Georg Simmel called, in 1900, the "calculating character of modern times," this earlier transition is its point of departure.[2]

It is not as simple as saying that to draw differently is to design differently or to build differently. But neither is that suggestion, which remains tacit in today's diatribes regarding the impact of computerization on architectural design, so far off that we should exchange it for the puerile fantasy that the architect independently decides, as a matter of creative will, the architecture. Witness the furious return of the rhetorical watercolor sketch or the shaky, hand-drawn line to the boardrooms of today's architectural discourse, as clients are time and again sold the story that "so-called man"—as the German media philosopher Friedrich Kittler used to say—is the master of (usually) his destiny, the Promethean creator of his world, by virtue of direct contact, through the magic wand of the pencil or the brush, with cosmic poiesis. If nothing else, the premise that different ways of drawing yield different architectural results has thereby been converted, in both the academic design studio and the professional marketplace, from a testable hypothesis to a master narrative with performative qualities all its own.

Seen in this light, design with computers is not so very different from design with charcoal, in that attaching message to medium presumes in both cases that architectural content is a linear function of its means of conveyance. Or, to put it differently, it may seem that drawing materials carry an ideological burden similar to building materials, wherein, say, charcoal is to brick what Autocad is to drywall. There is some logical consistency to this, but less in terms of semantics than in terms of pragmatics, if by pragmatics we understand more than simply use, or, in today's idiom, performance. To begin with, we must allow pragmatics or performativity to encompass the realm of nonlinear causality that constitutes the uncertain ground of constructionist models of knowledge. Constructionism can itself mean many things; here, let it simply mean the proposition that knowledge and cultural production, including technological production, are historically contingent. What any given discourse maintains as truth is not simply the opposite of untruth, but the outcome of a potentially (and numerically) infinite set of interacting historical factors—economic, technological, social, aesthetic—some of which, like modes of architectural visualization, can be isolated and studied in their variants, in order to determine the rules or principles that govern their interaction with other factors.

In focusing on techniques or material processes, such a view inherently favors the ontic over the ontological, or the realm of historically contingent facts over that of metaphysical absolutes.[3] In which light most dogma regarding the constitutive role of visual or technical media in architectural design and production rapidly discloses its metaphysical ambitions. That is, rather than contemplate the historical consequences of this or that technique, architects concerned with such matters tend to imply that this or that technique gives greater or lesser access to timeless truths. If what is on the table in our case is the technique or set of techniques that have been named "parametricism,"[4] my first concern is to evaluate its historicity before considering its claims to metaphysical authority.

But neither is constructionism, as I have described it, immune to the accusation that it substitutes a metaphysics of contingency for a metaphysics of the absolute. By regarding visualization and production techniques as subject to the vicissitudes of history, context, or culture rather than as pathways to universal value, some might say that we risk a relativism that, in the final analysis, remains unable to establish the priority of this or that set of techniques since all techniques are historically constructed and hence, derive from competing priorities that are, in theory, equally valid. As a result, we could be further accused of assigning to History, contingency's master, the status of indisputable sovereign. The first part of this critique is tangentially applicable to the claim made by postmodernists regarding the universal aspirations of Modernism, which many of its apologists saw as historically inevitable: that deference to local context, style, or tradition was ethically preferable to the imperious rule of number, standardization, and abstraction with which Modernism was associated.

The retort has often been that architectural pluralism or relativism imposes a tyranny of its own, willingly submitting to the consumerist spectacle by offering up architectural styles like so many commodities on supermarket shelves. In its broadest, most inclusive forms, so-called parametric architecture has sought to resolve the apparent antinomy that pits postmodernist variability against modernist calculability simply by subsuming one within the other. Its most recognizably (i.e., stylistically) postmodern instances are to be found in the design practices of commercial builders who use digital modeling and fabrication techniques to produce variations on traditional stair rails, banisters, and other components for suburban houses in place of the "abstract" geometrical patterns that are favored by their vanguardist colleagues.[5] But the underlying philosophical and political questions are the same. To address these questions, we must establish the basics of an intellectual history in which they become intelligible.

First of all, what is a parameter? Nearly every field uses the term: physics, economics, statistics, music, linguistics, medicine, psychology, genetics, computer science, philosophy, geology, climate science, literature, and, of course, architecture. This does not make parameter a universal concept; on the contrary, it relativizes it since many of these fields use the term in a specialized way. In general, we can divide the usage of "parameter" across fields into two classes: quantitative and qualitative. A parameter controlling or guiding a particular outcome could be numerical, as, say, in the pair of equations, $x = a \cos \theta$ and $y = b \sin \theta$, where θ runs between 0 and 2Π, which constitute the mathematical description of all points x, y lying on an ellipse, where the angle θ is the parameter.[6] All such parameters are quantitative by nature. But a parameter can also correspond to a certain quality such as, say, "redness," whether or not that quality is strictly measurable. In which case, the more "red" someone is, the more closely she may be thought to adhere to a particular political ideology, which in turn may affect her social standing, and so on.

But even here, my qualitative example (degrees of "redness") implies a certain quantitative aspect, which is expressible in terms like "more" or "less." This is because we cannot avoid mathematics, even in our ordinary or "natural" language. Lying within that language are numbers, or at least, numerical concepts. So even in its qualitative form, a parameter is a number, more or less.

This is not to say that all language is ultimately a branch of mathematics; it is simply to set aside any absolute distinction between quantitative and qualitative statements. But what, after all, is a number? This is an age-old philosophical question with quite a controversial history, of which it is worth reviewing a few excerpts. Typically, architects, like architectural theorists and historians (myself included), are not mathematicians. Still, they regularly make use of mathematical concepts that participate in the everydayness of number, a characteristic that is expressed, for example, in the fondness of mathematicians and logicians for illustrating complex ideas with examples drawn from everyday experiences. I will therefore reverse that habit and use everyday examples related to architecture to suggest ways in which associated concepts, like objecthood and color, give access to rudimentary problems in the philosophy of number that may help illuminate the more obscure corners of parametric thought.

In 1919, for example, Bertrand Russell, the British logician who helped to popularize the set theory developed by the German mathematician Georg Cantor, gave what remains a classic definition of number. In its simplest form, Russell defined number as "anything which is the number of some class," in which "the number of a class is all those things that are similar to it."[7] This means that the number two will be defined as the class of all sets containing two elements, or the class that contains every instance of pairing or coupling, whether these are pairs of houses or pairs of persons. In contrast to this set-theoretical definition, where number is ontology, or logical essence, Russell's student, Ludwig Wittgenstein, in his "late" period of the 1930s and early 1940s, redefined number as an element in a language game. By this he meant that there is no number "two" prior to its enunciation in language, as "two." More specifically, a number belongs to what Wittgenstein called the language game of "inventing a name for something," or the game of "ostensive definition." In such games, which obey strict rules, there is nevertheless always room for confusion. If I point to two houses and say "this is called two," you cannot be sure whether I am referring to all pairs in general or this particular pair of particular things in a particular place. To clarify, I might say, in the manner of set theory: "This number is called two," thereby fixing the term "number" grammatically as a denotative. Once fixed grammatically, however, the word "number" still requires other words to define it, which in turn require other words to define them, and so on.[8]

This does not mean that the meaning of "number" is utterly dependent on the self-referential abyss of language and is therefore undecidable; it only means that the meaning of "number" is context dependent; in other words, what we mean by "number" is an extra-mathematical function of its usage in a complex, socially enframed language game. On this view, there are no purely mathematical numbers, or no numbers outside language and the social relations it entails. Wittgenstein's redefinition therefore takes number out of the rarified, Platonic sphere of mathematical logic and puts it back in the world.

All of which becomes more immediately relevant to us when we recall that Wittgenstein's concept of language games was taken up in 1979 by the French philosopher Jean-François Lyotard in his influential "report on knowledge" titled *The Postmodern Condition*. Although it is mostly remembered for announcing the disappearance of the "grand narratives" that shaped the modern period, such as the narrative of progress or the narrative of emancipation, the main concern of Lyotard's little book was a shift in regimes of legitimation, or the determination of truth. This shift, Lyotard claimed, was proper to what he called a "computerized society," where, as knowledge ceased to be an end in itself and became commodified as a unit of networked exchange, assisted by massive governmental and corporate investment, the determination of truth, including scientific truth, increasingly became a matter of the political and economic power to decide what is true and what is not.[9]

Although many have interpreted this argument as hopelessly relativistic, it is not; it is, as Lyotard says, pragmatic (we could even say pragmatist) in the philosophical sense. That is, language games help to explain how different forms of truth-making work at a practical level. For this Lyotard draws on a series of sources in addition to Wittgenstein. These include the pragmatist semiotics initiated by the nineteenth-century American philosopher Charles Sanders Peirce and the speech act theory of the mid-century British philosopher J.L. Austin. Essentially, Lyotard argues that the truth-value of different types of statements depends on the social contract among speakers that defines the rules of the game they are playing. Examples include descriptive utterances ("this is true" because you and I agree that it is true, as when I say "this is an interesting lecture") and performative utterances ("this is true" because I am in a position to make it true, as when I say "this lecture is over"). According to Lyotard, such rules apply to scientific truth claims just as they do to everyday speech. And so, if we accept Wittgenstein's "turn" to language as an alternative to the ontologies of logical positivism, even the definition and usage of number is not determined by an absolute, timeless truth, as philosophers like Russell believed; rather, it is the consequence of socially constructed rules defining the language game of mathematics as these contribute to a general "agonistics of language" in which, as Lyotard says, "to speak is to fight."[10]

Lyotard's assessment of postmodernity further accepts the German social systems theorist Niklas Luhmann's proposition that, as Lyotard puts it, "in postindustrial [or computerized] societies the normativity of laws is replaced by the performativity of procedures."[11] As it happens, we have a local instance of this in architecture, where we can observe everywhere the normative "laws" of Euclidean geometry, construed by European humanists as absolute and God-given and taken up by their modernist successors as a sort of secular divine, being dismantled and replaced by computerized, rule-based procedures, such as the "design methods" devised by the 1960s Cambridge group or their rediscovery by today's parametricisms. In such games, which could, strictly speaking, be described as language games (think for example of the language of command in computer code), legitimacy is defined as obedience to quasi-bureaucratic, rule-based procedure, yielding statements of the type: "My design is beautiful not because it obeys the laws of absolute proportion, but because it was generated in such-and-such a manner, according to such-and-such set of rules."

Lyotard calls this "legitimation by power" since the legitimacy of the rules or procedures themselves is determined by the performativity criterion, not necessarily in the sense of optimizing some function, but in the sense of the performative utterance.[12] Such an utterance is self-legitimating: this particular set of procedures is optimal because I am, or the system of which I am a function, in a position to determine it to be so. Therefore I, or the system of which I am a function, allocate further research funds to this set of procedures on the basis of this determination and cancel the funding allocated to other utterances that fail to meet the self-legitimating performativity criterion. Such a sequence has nothing to do with either truth-value or use-value in any absolute sense. It has only to do with the capacity to play the game according to the rules that are in place, or to change those rules to favor certain outcomes.

Lyotard's account of the pragmatics of knowledge within postmodern language games learned much from Luhmann's adaptation of the mid-century American sociologist Talcott Parsons's functionalist model of society as a system of systems. But Lyotard argues that Luhmann's notion of the self-referential, self-organizing social system builds into itself the administration of consensus, which is related to the fetishization of procedure as the sole legitimation game in postindustrial societies. In Lyotard's view, the system or language game of science, for example, is programmed to reject (or to defund) those results that constitute genuinely new moves or that require a change in the game's basic rules. Translated to social relations, this suggests conformity (something like Thomas Kuhn's "normal science") akin to the totally administered lifeworld feared by critics of corporate life. It also bears comparison to the Frankfurt School philosophers Theodor Adorno and Max Horkheimer's merciless critique of Hollywood's "culture industry," which they describe as a dream factory relentlessly producing identification with the system and a desire for what that system desires, in a perfectly coherent instance of autopoietic self-organization and self-regulation.

Lyotard calls this administrative violence "terror," or, in his words, "the efficiency gained by eliminating, or threatening to eliminate, a player from the language game one shares with him."[13] Hence, the politics of knowledge becomes, fundamentally, a question of language. Although Lyotard himself cautioned against architects "getting rid of the Bauhaus project" by "throwing out the baby of experimentation with the bathwater of functionalism,"[14] those very architects, whether we call them postmodernists or not, are the inheritors of modernity's "linguistic turn," which runs from Wittgenstein to Lyotard and well beyond. But language is not the only domain in which the singular authority of number has met its own contingency, or variability.

Lyotard's former colleague at the journal *Socialisme ou Barbarie* (Socialism or Barbarism), the Greek-French philosopher Cornelius Castoriadis, for instance, was a reader and friend of the Chilean theoretical biologist Francisco Varela, from whose work on the self-generation of biological systems he drew strong analogies with his own political philosophy of radical democratic or "autonomous" self-government. Like Luhmann, Castoriadis borrowed the term "autopoiesis," or self-creation, from Varela. But Castoriadis used the term somewhat differently, to designate the fundamental creativity he ascribed to social beings, or more specifically, to what he called the "social imaginary." For Castoriadis, the "imaginary" was not a Lacanian speculum, or a deformed looking-glass image of an unrepresentable reality, nor was it simply a linguistic construct; it was both the product and the producer of lived realities, including what he called the "imaginary institution of society" itself.

To explain, Castoriadis proposed starting with what he called the "banal facts," namely, that

> There is no society without arithmetic. There is no society without myth. In today's society, arithmetic is, of course, one of the main myths. There is not and cannot be a "rational" basis for the dominance of quantification in contemporary society. Quantification is merely the expression of one of its dominant imaginary significations: whatever cannot be counted does not exist.[15]

Accordingly, our society's dominant imaginary significations occupy two related dimensions: what Castoriadis called the "ensemblistic-identitary" (or "set theoretical, logical," calculating, computational) dimension and the "properly imaginary dimension." Like Wittgenstein, Castoriadis wanted to pry open the ontological closure of set theory, although in this case not with language but with life. In a 1995 exchange with the biologist Varela on French radio, Castoriadis associated the set-theoretical or "identitary" dimension with a computational model of human cognition, to which he opposed the irreducibly "creative" or autopoietic dimension of the "instituting imaginary." By which he meant an imaginary dimension of being that incessantly produces itself and in so doing, produces a world "for-itself." Castoriadis, who was a student of the phenomenologist Maurice Merleau-Ponty (also a source for Varela), frequently returned to the example of color. In his view, color does not exist independently of its instituting imaginaries. Certain living beings perceive color, but, as Castoriadis put it, "the world [or, we could say, the instituting imaginary] of the physicist does not have color; it has wavelengths."[16] So, translated to our context, the parameter of "redness," which we have already recognized as possessing a quantitative and a qualitative dimension, can be described in two ways: first, as a mathematical function that is expressed, say, in a Photoshop slider or color wheel, under which lies a pattern of ones and zeroes; and second, as the effect of an instituting imaginary, or a combination of interacting imaginaries, for which political redness may or may not coincide with the color of a flag, the color of a pixel, or the color of a book.

Red or not, philosophers hardly agree on the political or proto-political characteristics of number. Alain Badiou, for one, has dedicated an entire book—if not an entire philosophy—to reversing Wittgenstein's move from logic to language, not by positing an extra-logical or extra-mathematical order of being (as does Castoriadis) but by returning to mathematics, and in particular, to set theory, in order to locate what he calls "another idea of number" in its irreducible rupture, or "event."[17] Like Castoriadis, Lyotard, and many other philosophers who in different ways stress event-like indeterminacies or multiplicities over closed or internally homogeneous sets, Badiou does so to combat what he calls "number's despotism," which circumscribes our notion of the political by reducing politics (and economics) to instrumental acts of counting.[18] But unlike these others, Badiou asserts number's ontology, over and above its ontics, as the basis for resisting or overcoming such despotism.

Badiou also proposes "mathematics as ontology" over and against the historicism that has been implicit thus far in my emphasis on situational contingency, be it Lyotard's or Wittgenstein's pragmatics of language, or the collective, socially constituted imaginaries of Castoriadis. In this respect and in general, Badiou's philosophy is anticonstructionist, without necessarily falling into the abyss of a mystical (or, as he would put it, religious) essentialism. That is because his ontology of number is not based, as he again might say, in the oneness of the One; it is founded on multiplicities to infinity. These are not limited to simple multiples, or to serially infinite sets of ordinal numbers. Nor are they solely concerned with limit cases, or with the possibility that a seemingly infinite series will double up on itself, and repeat. Badiou's multiplicities, and ultimately, his appeal to the infinite, appear in the gap between numbers, in the simple step from one to two, as an unfathomable but utterly real event.

Badiou derives these arguments from the logic of sets by moving through some of the landmark hypotheses in its modern history and challenging or extending their conclusions. Others, equipped with specialized knowledge, have evaluated his tendency to metaphorize or anthropomorphize set theory's language and hence, distort its translation to human affairs by overstepping set theory's limits.[19] We might temporarily minimize this concern through recourse to one of Badiou's nonmathematical examples: the seriality of modern art. In his lofty meditation on the artistic and political life of the twentieth century (almost exclusively in Europe), titled simply *The Century*, Badiou provides an example of what he calls, elsewhere, the "grand style" of philosophizing that rejects historicism and epistemological niceties (the "little style") in favor of the "glacial antihumanism" of ontology and truth.[20] Announcing his method, Badiou declares: "For us philosophers, the question is not what took place in the century, but what was thought in it."[21] In response to which we learn that, among other things, the European twentieth century sought to think, repeatedly and in unique ways, the infinite. It did so, among other ways, in the negative, through the serial artwork—as in Duchamp's Readymades (the set of all urinals, etc.)—which makes visible "the power of the finite," as repetition.[22] Extended logically, such repetition tends toward infinity not because its series is endless, but because each repetition is an act, like the artistic construction of "situations" or happenings that followed in the modernist lineage devoted to the subsumption of art into life. "Ideally," says Badiou, "the twentieth century artwork is nothing other than the visibility of its own act." This visibility, which is finite, overcomes the "romantic pathos" of the artist-seer and of the singular artwork, "because the artwork has nothing infinite to show, save for its own active finitude."[23] All art is therefore in some sense choreography, "an art of formalization [of the set, of the series] rather than of the work." And formalization is "the great unifying power behind all the century's undertakings—from mathematics (formal logics) to politics ... by way of art ... ."[24]

Badiou's political philosophy, in which the paradigmatic event is revolution itself, and his philosophical poetics both depend on the transitivity (in the set-theoretical sense) of number from one realm to another, not as sign, but as being. The opposite of this is the case with the ideology—for we cannot rightly call it a thought, or even a theory—that architects call "parametricism." The parametricist premise is simple enough: being is multiple rather than one; reconceived as number, architecture transitively reproduces that multiplicity. But—and here is a crucial difference—not as an act that, in its repeatability and its ultimate sameness, indirectly reveals the inhuman infinite that lies between its numbered series. Rather, the aestheticized manipulation of parameters aims to domesticate multiplicity by saying, most definitively: Stop! And then: Repeat! Its affect is thus built around the problem of deciding which of the innumerable variables should be preferred. This problem above all others commands our attention as fields of parameters are deployed across the cityscape. We understand well that, in all such cases, the particular version selected (and perhaps built) is not in any real sense optimal; it is not and can never be, even for the most ardent systems modeler, a coordinated balance of variables that "solves" the problem of human habitation and formal expression once and for all. Rather, what matters is that the selected option appears—and, in effect, is made to appear—as simply one variation among many at which the serial process happens to have paused, and therefore, one among many instantiations of the set of equations by which that problem has been posed.

As in finance or cyberwar, the equations, rather than the value of this or that parameter, govern. Our society is saturated with such equations. We are surrounded by sliders, by pseudo-optimized curves, and by mini-max functions writing the code for the new capitalist organicism. It is an organicism rather than a "glacial antihumanism" because it encodes a nightmarishly emergent, autopoietic system that is fiendishly "alive" even as it shouts "Stop! Repeat!" to the calculations (and the calculators, both human and electronic) that underlie it. The artistic formulas of parametricism display these calculations in a finite being, such that the unruliness of number might finally be domesticated, and the attributes of that being—a building, a city, a chair, a collective—might finally be counted, rather than thought.

This is the dominant imaginary in which we find ourselves today: Badiou's intransigence notwithstanding, it is a self-perpetuating, performative reality, in which life itself is subject to number's "despotism," or the tyranny exerted by numbers when conceived as instruments rather than thoughts. Badiou enunciates the society's categorical imperative: "Count!"[25] To which we might add the question of who and what counts, in both senses. Namely, who is doing the counting and who is counted, in or out. Or, translated to the parametric idiom, what or who counts as a parameter and what or who does not, according to what rule of what numbers game—meaning, what language game—or, if you prefer, what form of life.

But politics is not counting, whether of votes, opinions, or anything else, nor is it the enumeration of statistics or the provision of choices. Nor is counting inherently political, in the strong sense of giving form to the *polis*. It is, however, an act. Whether it is a repressive act or an emancipatory one depends on how we respond to the disputes that I have outlined. It is not my aim to resolve them here, less still, to reduce their depth to the few generalizations that a sketch of this sort permits. Instead, I hope only to have shown what it would mean even to begin the necessary task of theorizing number in artistic, or architectural, thought.

In his meditations on the twentieth century, Badiou gives us a suggestive, if simplistic, glimpse of the antihumanistic aspirations of number, which he likens to the "Bauhaus in architecture: a building that renders nothing in particular, for it is reduced to a translucent, universally recognizable functionality; the kind of functionality that has forgotten every instance of stylistic particularity."[26] If there is something familiar in this, it is in the sense that still today architects and theorists of architecture resist that antihumanism as they return to a longstanding problem associated with the "Bauhaus in architecture," and with Modernism more generally. Namely, the problem of transforming humanity rather than simply counting its variants, as postmodernists tend to do.

The style called "parametricism" is only the latest entry in the list of efforts to discredit the project of transforming humanity, in this case by pretending to take up its numerical idiom. Parametric commands—"Stop! Repeat!"—run mathematical functions that replace the repetition of the same with the repetition of difference. But as we have seen, repetition does not necessarily extend the limit to infinity; it simply rehearses the act of counting. Any "ism" that pretends to replace the sameness of that act with its variability dons the sentimental cloak of the human—the one who counts, the calculator, the computer—to protect itself from the glacial stare of number. This cloak secures for the architect admission into the parametric dreams of capital, which, far from dehumanizing, depend on the warm reassurance that all the world is a bar graph, a fluctuation, a line. That this may in fact be true, though in a manner unthinkable to the managers, the counters, and the calculators, is a possibility we will have to leave for another day.

Notes

1. Hyungmin Pai, *The Portfolio and the Diagram: Architecture, Discourse, and Modernity in America* (Cambridge, MA: MIT Press, 2002).
2. Georg Simmel, *The Philosophy of Money*, ed. David Frisby, trans. Tom Bottomore and David Frisby, 2nd ed. (London: Routledge, 1990), p. 443.
3. Bernhard Siegert, "Cultural Techniques: Or the End of the Intellectual Postwar Era in German Media Theory," *Theory, Culture & Society* Vol. 30, No. 6 (2013), p. 57.
4. Patrik Schumacher, *The Autopoiesis of Architecture: A New Framework for Architecture*, Vols. 1 and 2 (Chichester: John Wiley & Sons, 2011).
5. The industry-wide application of digital design and fabrication techniques is discussed in Peggy Deamer & Phillip G. Bernstein, eds., *Building (in) the Future: Recasting Labor in Architecture* (New York: Princeton Architectural Press; New Haven: Yale School of Architecture, 2010).
6. This example and others like it are commonly given to define parameterization, as in *The Concise Oxford Dictionary of Mathematics*, 4th ed. (Oxford: Oxford University Press, 2009), online at http://www.oxfordreference.com/view/10.1093/acref/9780199235940.001.0001/acref-9780199235940-e-2105# [accessed January 2, 2015]. The Wikipedia entry under "parametric equation" gives a similar example, at http://en.wikipedia.org/wiki/Parametric_equation [accessed January 2, 2015].
7. Bertrand Russell, *Introduction to Mathematical Philosophy*, 2nd ed. [1920] (New York: Dover, 1993), pp. 18–19.
8. Ludwig Wittgenstein, *Philosophical Investigations*, trans. G.E.M. Anscombe (Oxford: Basil Blackwell, 1963), pp. 5e–15e.
9. Jean-François Lyotard, *The Postmodern Condition: A Report on Knowledge*, trans. Geoffrey Bennington (Minneapolis: University of Minnesota Press, 1984), p. 8.
10. Lyotard, *The Postmodern Condition*, p. 10.
11. Ibid., p. 46.
12. Ibid., p. 47.
13. Ibid., p. 63.
14. Jean-François Lyotard, "Answering the Question: What Is Postmodernism?," trans. Régis Durand, in *The Postmodern Condition*, p. 71.
15. Cornelius Castoriadis, "The Imaginary," in *World in Fragments: Writings on Politics, Society, Psychoanalysis, and the Imagination*, ed. and trans. David Ames Curtis (Stanford: Stanford University Press, 1997), p. 11.

16. "Life and Creation," Cornelius Castoriadis in dialogue with Francisco Varela, trans. John V. Garner, in Castoriadis, *Postscript on Insignificance: Dialogues with Cornelius Castoriadis*, ed. Gabriel Rockhill (New York: Continuum, 2001), p. 60. Interview from radio broadcast France Culture, 1995.
17. Alain Badiou, *Number and Numbers*, trans. Robin Mackay (Malden, MA: Polity Press, 2008), p. 4. On number as event, or "evental trans-being," p. 214.
18. Ibid., pp. 1–2.
19. Ricardo L. Nirenberg & David Nirenberg, "Badiou's Number: A Critique of Mathematics as Ontology," *Critical Inquiry* Vol. 37, No. 4 (Summer 2011), pp. 583–614.
20. Alain Badiou, *The Century*, trans. Alberto Toscano (Malden, MA: Polity Press, 2007), and Badiou, "Mathematics and Philosophy: The Grand Style and the Little Style," in *Theoretical Writings*, ed. and trans. Ray Brassier and Alberto Toscano (New York: Continuum, 2006), pp. 3–21.
21. Ibid., p. 3.
22. Ibid., p. 158.
23. Ibid., p. 159.
24. Ibid., p. 160.
25. Badiou, *Number and Numbers*, p. 1.
26. Badiou, *The Century*, p. 161.

Chapter 4

There is no such thing as political architecture. There is no such thing as digital architecture

Neil Leach

This chapter argues that there is no such thing as political architecture, and equally no such thing as digital architecture. It argues that politics is most properly defined as a *praxis* or process, and that it therefore does not reside in the "order of things" (Foucault). Likewise it argues that the term "digital"—as distinct from "analog"—is an adjective used most properly to refer to tools, and that therefore the digital domain must also be understood in terms of processes where these tools are deployed. As such, it argues that neither of the terms, "political" or "digital," refers to objects—such as buildings—but rather to practices/processes. Finally, it proposes a "theory of affordances" that might be used to understand the hegemonic effects that may result from the use of certain tools and practices/processes, even though those tools and practices/

The distinction is made here between the term "architecture"—to refer to buildings in general—and the term "Architecture" commonly used to refer to the "discourse of architecture." The title of this chapter is deliberately provocative. It is inspired by the title of one of Stanley Fish's books, *There's No Such Thing as Free Speech – And It's a Good Thing Too*. For many it might seem perverse that I am claiming that there is no such thing as political architecture and no such thing as digital architecture. After all, everyone wants to believe that their own field is highly meaningful, and has deeply political consequences. Such a claim would therefore seem to undermine much of what they believe in. However, I am stating what seems to me to be irrevocably the case after several years of reflection and writing in the fields of both the politics of space and digital design. But let me clarify one point: I do not wish to associate myself with those individuals who claim that the realm of digital design—unlike that of analog design—is lacking substance, and has no political, social, or economic efficacy. Such thinking seems to stem from a misunderstanding of both digital and analog design, and yet it is surprisingly common. Whatever we might call digital design has no more or no less political, social, and economic efficacy than any other kind of design. Yet we must remain alert to unsubstantiated claims about the significance of any discipline.

processes are not responsible—in and of themselves—for any particular style or aesthetic expression.

There is no such thing as political architecture

Figure 4.1 House of the People, Bucharest. Image available at http://commons.wikimedia .org/wiki/File:Largest_structure_in_Europe_and_second_in_the_World_after_The_Pentagon_-_ Palace_of_the_Parliament_%2811321680666%29.jpg under a Creative Commons Attribution 2.0 Generic license.

In 1995, I stood in front of Nicolae Ceauşescu's as yet unfinished palace in Bucharest, now known as the House of the People. The building had caused outrage among architects, largely because of its kitsch postmodern style of architecture. Most architects, it would seem, wanted to see the building demolished.[1] At the time much discussion about the building—among architects at least—seemed to center on its appearance or aesthetic expression. Some architectural historians even seemed to take the view that if only Ceauşescu had commissioned a better design from a better architect, the result would have been perfectly fine, despite the controversy that surrounded Ceauşescu himself.

Standing next to me at the time was the American cultural theorist Fredric Jameson. His view was quite different. "There's nothing wrong with the building as such," I recall him saying. "It might be an example of bad architecture.

But from a political perspective there is nothing actually wrong with it. What is wrong, however, is that—in order to build it—so many people were evicted from their homes, and the economic resources of a nation were drained." Jameson's comments threw into perspective the widespread opinion among architects at the time that the building seemingly tainted because it was badly designed, although for others the building was tainted through its association with Ceaușescu.

Let us overlook for the moment the fact that the building has been seemingly reappropriated as a popular iconic structure at the center of the most expensive area of real estate in Bucharest, and effectively rinsed of its negative associations with Ceaușescu. Instead, let us consider the actual relationship between architecture and politics. Let me start, then, with a question. If there were such a thing as a political building, what would it be? We might follow this up, perhaps, with a further question: how might we define the political?

Within the popular imagination there has been little doubt about architecture's capacity to condition a response within the user. Indeed, the common view seems to be encapsulated in Georges Bataille's definition of architecture. For Bataille, architecture—especially monumental architecture—not only reflects the politics of an epoch, but also has a marked influence on the social.

> Architecture is the expression of the true nature of society, as physiognomy is the expression of the nature of the individual. However, this comparison is applicable, above all, to the physiognomy of officials (prelates, magistrates, admirals). In fact, only society's ideal nature—that of authoritative command and prohibition—expresses itself in actual architectural constructions. Thus great monuments rise up like dams, opposing a logic of majesty and authority on all unquiet elements; it is in the form of cathedrals and palaces that the church and state speak to and impose silence upon the crowds.[2]

Equally, architects themselves tend to believe in the political and social efficacy of their designs. *Architecture ou Révolution*, wrote Le Corbusier in 1922. "It is the question of building which lies at the root of the social unrest of today; architecture or revolution."[3] Le Corbusier, in common with many architects of the Modern Movement, was convinced of the social role of architecture. In an era of great social and political change, Le Corbusier perceived architecture as a crucial instrument in addressing the ills of contemporary society. An appropriate architecture would combat social unrest. Architecture could prevent revolution.

While Le Corbusier saw architecture as a way of avoiding revolution, the architects of postrevolutionary Russia saw architecture as a way of supporting the aims and ideals of a Marxist revolution. Architectural theorists, such as Alexei Gan and Moisei Ginzburg, looked to architecture for a means of resolving the particular problems of postrevolutionary Marxist society. Buildings should not simply reflect passively changing social conditions; they should be active instruments of change. Thus, for Gan and Ginzburg buildings themselves were to be "revolutionary," and were to operate as active social condensers.[4]

On the face of it, Le Corbusier's position seems diametrically opposed to that of Gan and Ginzburg. Yet, an alternative reading is possible, and it could be argued that Le Corbusier spoke of avoiding political "revolution" not because he was opposed to the concept of revolution as such, but rather because he recognized in architecture the possibility of a "revolution" that would go beyond the political. As Fredric Jameson has observed, "he saw the construction and the constitution of new spaces as the most revolutionary act, and one that could 'replace' the narrowly political revolution of the mere seizure of power."[5] Thus, far from being against revolution, Le Corbusier could be seen as a supporter of reform in its most radical and far-reaching sense. It is clear that both Le Corbusier and the architects of the new Russia recognized in architecture the same potential, the possibility of alleviating social problems and of creating a new and better world. Architecture for the pioneers of the Modern Movement had a role as a democratic force within a democratic society. Architecture was to be a force of liberation, overtly political and emancipatory in its outlook.

The relationship between architecture and revolution needs to be interrogated beyond the naive utopianism of the Modern Movement. The term "revolution" should not be taken lightly, nor treated uncritically. Too easily such a term may be appropriated to dress up shifts in political power, which, far from overturning a previous regime, simply replicate the *status quo* in an alternative formal variant. Too easily, also, such a term may be smuggled into empty commercial slogans to refer to merely ephemeral changes in fashion.

What influence, then, can architecture claim to have on the social and the political? What is the status of architecture as a force of social change? What is the link between aesthetics and politics? What relationship may there be between architecture and revolution? Indeed can there even be a "revolutionary" architecture?

The interaction between architecture and the political deserves to be interrogated further. This is not to deny, of course, the status of building as a political act. Certainly, if we are to believe Stanley Fish, every act—and this includes building—is inscribed within some ideological position.[6] There is no platform, according to Fish, which is not constrained by some ideological imperative.

Indeed, there needs to be an ideological content in that this is precisely what gives an act its force. This may not be obvious because ideology remains largely invisible, yet it is through its very invisibility that ideology derives its potential. Ideology constitutes a form of background level of consciousness that influences all our actions.

A distinction must be made, however, between the process of constructing a building and subsequent semantic "readings" of that building. The political content of the process of construction is perhaps the more obvious, but it is likewise the more often overlooked and forgotten. In the case of the Stalinallee in Berlin, for example, the process of construction was deeply political and was marked by considerable social unrest. Demonstrations over the low level of pay for building workers on the project erupted on June 16, 1953, and spread the following day to other parts of the city.[7] As could be expected, the demonstrations were brutally suppressed, and several demonstrators were killed. Yet, what dominates discussion of the Stalinallee is not this all-but-forgotten moment in its construction, but the question of whether the project can be read semantically as "totalitarian."[8]

It is precisely in these semantic readings of architecture that the fragility of associations between architecture and the political becomes most apparent. In their discussion of "democratic" architecture, Charles Jencks and Maggie Valentine recognize the subject as problematic. They observe that neither Frank Lloyd Wright nor Vincent Scully, both of whom had written on the subject of architecture and democracy, had managed to relate the politics to any typology or style of building.[9] Yet, while they also note that Aldo Rossi and others had claimed that there was no direct link between style and politics, they themselves persist in an attempt to define an "architecture of democracy." Their approach relies on semantic readings. For Jencks and Valentine, as it transpires, the problem rests ultimately in the complex "codes" that "democratic architecture" adopts. It must avoid excessive uniformity ("an architecture of democracy that is uniform is as absurd as a democracy of identical citizens") yet equally it should avoid excessive variety ("an architecture where every building is in a different style is as privatized as a megalopolis of consumers"). "Thus a democratic style," they conclude, "(…) is at once shared, abstract, individualized and disharmonious."[10] Jencks and Valentine emphasize the aesthetic dimension, as though this has some direct bearing on the political. Yet, their argument is undone by its own internal inconsistencies. How can classical architecture symbolize both Greek democracy and Italian fascism? Can there be any essential politics to a style of architecture? Can there ever be a "democratic architecture"?

More recently, architectural interventions, such as the remodeling of the Reichstag, Berlin, by Foster + Partners, are seen as deeply political because the transparency of the glass architecture suggests a certain transparency in the political processes taking place there. But is this physical transparency of the glass not merely *symbolic* of the political processes? And, does it have any actual political impact on the operations of government itself?

Here, we must recognize that political content in architecture must be seen as associative. Architecture can only be imbued with political content through a process of "mapping." Architecture achieves its political—and hence equally its gendered—status through semantic associations, which exist within a temporal framework and are inherently unstable. These semantic associations depend on an historical memory within the collective imagination. Once this memory fades, the semantic associations will be lost, and the building may be reappropriated according to new ideological imperatives. Thus, the pyramids' emblems, no doubt, of totalitarian rule to the slaves who built them, have now shifted their symbolic content to icons of tourism. A similar process inevitably occurs when a building changes its use, from Victorian villa to academic department, from police station to brothel, from dictator's palace to casino. Unless the memory of its previous social use is retained, all earlier associations are erased. While a building through its associations might appear as deeply political, it must be understood that these politics are not an attribute of the architectural form itself. Political content does not reside in architectural form. It is merely grafted on to it by a process that is strictly allegorical. To perceive the political meaning one has to understand the allegorical system in which it is encoded. Yet, this is not the allegorical system that one might identify, for example, with Renaissance painting, where allegory relies on a narrative of fixed symbols with which the painter works. The allegory to which I refer is an allegory of association. A closer comparison, therefore, might be the way in which abstract painting has been read as political, and promoted by the Central Intelligence Agency (CIA)—so the story goes—as a tool of postwar propaganda.

Fredric Jameson highlights the problem of the allegorical nature of this "mapping" of the political onto the architectural. Whatever political content might seem to be invested in architectural form may subsequently be erased or rewritten:

> I have come to think that no work of art or culture can set out to be political once and for all, no matter how ostentatiously it labels itself as such, for there can never be any guarantee that it will be used the way it demands. A great political art (Brecht) can be taken as a pure and apolitical art; art that seems to want to be merely aesthetic and decorative can be rewritten as political with

energetic interpretation. The political rewriting or appropriation, then, the political use, must be allegorical; you have to know that this is, what it is supposed to be or mean—in itself it is inert.[11]

He further elaborates this in his incisive critique of Kenneth Frampton's essay on critical regionalism. What is crucial is the "social ground" of architecture. When removed from its contextual situation, architectural form would be exposed for what it is. Architectural form, as Jameson notes, "would lack all political and allegorical efficacy" once taken out of the social and cultural movements that lend it this force. This is not to deny that architecture may indeed have "political and allegorical efficacy," but rather to recognize that it merely serves as a vehicle for this within a given "social ground." Thus, to return to our earlier quotations from Bataille, it is "in the *form* of"—through the *medium* of—"cathedrals and palaces that the church and state speak to and impose silence upon the crowds." Remove the memory of the church and state, and the buildings would become empty vessels to be appropriated toward some other political end. Yet, the point to be made here is that architecture is *always* contextualized within some social ground. It is therefore always appropriated toward some political end. But that political content is *not* a property of the architectural form itself. To view architectural form as inherently "politicized" is, for Jameson, a misguided project:

> It was one of the signal errors of the artistic activism of the 1960s to suppose that there existed, in advance, forms that were in and of themselves endowed with a political, and even revolutionary, potential by virtue of their own intrinsic properties.[12]

Architecture, then, may be seen to be the product of political and social forces, yet, once built, any political reading of it must be allegorical. As such, we should take care to distinguish an aesthetic reading of form from a political reading of content, even though the aesthetic terminology—"reactionary," "totalitarian," etc.—may ape the political. Failure to recognize this distinction would allow the difference between the two to be elided, and the aesthetic to be read as necessarily political.

Indeed, the shortcomings of any attempt to "read" a politics into architectural form are brought out by the contradictions that may exist between such "readings" and the practices that actually take place within the building. The importance of the consideration of practice over semantic concerns has been highlighted by Adrian Rifkin in the context of Jean Nouvel's *Institut du Monde Arabe*.[13] This is a building that purports to celebrate "arabness" through the arabesque patterning of the facade. Yet, if we focus less on semantic readings of the facade and more on a politics of use of the building itself, we may discover that—far from celebrating "arabness"—the building replicates the cultural imperialism that is at play elsewhere in Paris. While the elegant Parisians eat their couscous in the restaurants, the Arabs themselves may be seen working in the kitchens. In short, the building is supporting, rather than resisting, the dominant "Orientalizing," cultural impulse. All this begins to call into question not only the process of reading a politics into architectural form, but also the effect that any such political reading might have on the user.

If then we are looking for an example of how we associate architecture with politics, we should look perhaps to the work of Krysztof Wodiczko, who projects politically loaded images on to buildings as a technique for making us think what that building stands for. Famously, for example, he projected a swastika on to South Africa House in London during the apartheid regime in order to express the fact that South Africa at the time subscribed to a "fascist" politics. Equally, we "project" associations on to architecture when we associate a building with a particular politics. But there is nothing intrinsic to the architecture that retains that association. The Union Buildings in Pretoria, for example, which likewise might have been associated with a fascist politics during the time of apartheid, would now be associated with a more liberal postapartheid politics. Indeed, as Nelson Mandela walked across the threshold of that building, he effectively recoded it and reappropriated it for a more enlightened form of government.

Figure 4.2 Krysztof Wodiczko, projected image of a swastika on South Africa House, London, 1985. Photo reproduced with permission from Krysztof Wodiczko.

The use of space can therefore be political, even if the aesthetic cannot be. Yet, one might still argue that architecture—in its very physical form—must indeed be political, through the influence that it exerts on the users of a building. In other words, there is an association to be made between the form of a space and the political *praxis* afforded within that space. This prompts the further question as to whether architecture in its physical form may somehow influence the politics of use.

Space, knowledge and power

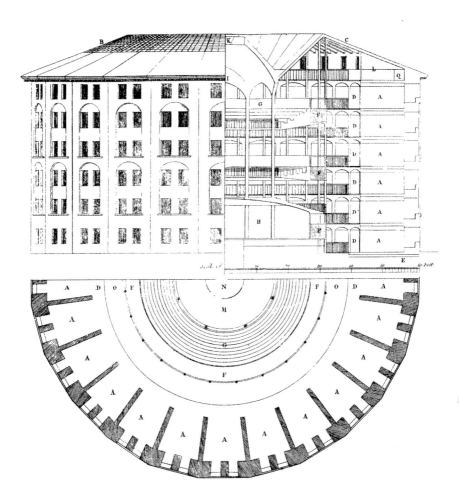

Figure 4.3 Elevation, Plan and Section of Jeremy Bentham's Panopticon penitentiary, drawn by Willey Reveley, 1791.

One of the central preoccupations for Michel Foucault is the relationship between power and space, and he throws some light on this issue in his discussion of Jeremy Bentham's panopticon. In this now famous piece, Foucault explores the question of how architectural form may influence social behavior. The panopticon is a plan for a prison. It has a central tower in which the guard sits, and the cells are arranged radially, so that from the tower the guard is afforded a view all around—as the name "panopticon" implies—into each of the cells. Meanwhile, the openings in the tower itself, through blinds and other devices, prevent the inmates in the cells from knowing whether or not the guard is looking at them. Thus, the inmates remain under the perpetual control of the gaze of the guard.[14] The principle that Foucault is trying to illustrate is that the architecture may become an apparatus "for creating and sustaining a power relationship independent of the person who operates it."[15] In other words, it is the architectural form of the panopticon that helps to engender a form of social control. Such an example would seem to suggest the possibility of architecture determining social behavior.

Foucault revisits this question in a subsequent interview with Paul Rabinow, where he acknowledges that architects are not necessarily "the masters of space" that they once were, or believed themselves to be.[16] Thus, he appears to qualify this position on the capacity for architecture to determine social behavior. On the question of whether there could be an architecture that would act as a force of either liberation or oppression, Foucault concludes that "liberation" and "oppression" are not mutually exclusive, and that even in that most oppressive of structures, some form of "resistance" may be in operation. Liberty, for Foucault, is a practice that cannot be "established by the project itself." "The liberty of men is never assured by the institutions and laws that are intended to guarantee them."[17]

Architecture, therefore, cannot in itself be liberative or repressive. As Foucault comments, "I think that it can never be inherent in the structure of things to guarantee the exercise of freedom. The guarantee of freedom is freedom."[18] Architectural form, Foucault concludes, cannot in itself resolve social problems. It is only politics that can address them, although architecture can contribute in some way, provided it is in league with the political. Thus Foucault concludes: "I think that [architecture] can and does produce positive effects when the liberating intentions of the architect coincide with the real practice of people in the exercise of their freedom."[19] Foucault is therefore not contradicting but merely qualifying his earlier comments on the panopticon. It is not the form of the panopticon that controls the behavior of the inmates. Rather, it is the politics of use—the fact that the building is operating as a prison—that is ultimately determinant of behavior, and the architecture is merely supporting that politics of use through its efficient layout.

The position of Michel Foucault on this matter is clear. In opposition to the utopian visions of Le Corbusier and others, Foucault would emphasize the politics of everyday life over architectural form as the principal determinant of social behavior. "The architect," he comments, "has no power over me."[20] According to such an approach, there could be no "revolutionary" architecture in the Marcusian sense of an architecture that might constitute some critical force of change. Yet, this is not to deny the capacity for architecture to "produce positive effects" when it is in league with the practice of politics. Such an approach, of course, introduces an important temporal dimension into consideration. As political practice changes, so the efficacy of the architectural form to support that practice may itself be compromised.

We might therefore surmise that physical form matters relatively little. It is a question of how a building is used, and that the political associations of a building are simply a reflection of that use. The Architectural Association, for example, occupies a series of terraced properties in London—34-38 Bedford Square. Because of the politics of use of those properties, we have come to associate them with a school of architecture. But the properties in Bedford Square were designed originally as family dwellings, although most have now become offices. Thus, although these particular properties have been associated with the activities of a school of architecture, they might equally have been used for other activities. For example, they might have become a police station, a brothel or a crèche for the children of single mothers. Indeed, here we might follow the thinking of Ludwig Wittgenstein on language. If we want to understand what a word means, we should not consult a dictionary or refer to books on literary theory. Rather, we should simply see how that word is being used in a particular context. For example, if I was playing a game of tennis, and called my partner a "bastard" when s/he managed to hit a particularly difficult shot that I was unable to return, it is not that I am suggesting that my opponent is illegitimate. Rather, the term "bastard" is part of a familiar lexicon of words used in such contexts, which gain their meaning from their context. The same is true for architecture. It gains its meaning from the activities that take place there, that haunt it "like a ghost."

Architecture, then, is traditionally seen as built politics, yet the problem is considerably more complex than might first appear. Extrapolating from Foucault's argument, we might conclude that there is nothing inherently political about any building or any style of architecture. It is a question rather of what political associations a building may have. Buildings, according to the logic of Foucault's argument, would have no inherent politics, if by "politics" we infer a capacity to influence the social. Rather, a building may facilitate—to a greater or lesser extent—the practice of those politics through its very physical form. We may recognize, for example, the naivety of the Jeffersonian "grid-iron" plan, which was carpeted across the United States in an effort to promote democracy. The supposed democracy of an antihierarchical, uniform layout such as the grid was of course challenged by the use of that form in the layout of that most antidemocratic of spaces, the concentration camp.

It is only perhaps if we are to understand architecture, along with the other visual arts, as offering a form of backdrop against which to forge some new political identity that we might recognize a political role for architecture, albeit indirect. For this backdrop, although neutral in itself, will always have some political "content" projected on to it. And it is as a "political backdrop"—politicized, that is, in the eyes of the population—that the architecture can act as a form of screen "reflecting" certain political values. As it is "encoded" in this way, the building will be seen to embody that new national identity. And, it is precisely through the population reading itself into this "screen" as though it were a mirror that a new sense of national identity might be forged.

There is no such thing as digital architecture

For Patrik Schumacher, there is a new global style of architecture—one that he calls "parametricism."

> There is a global convergence in recent avant-garde architecture that justifies its designation as a new style: parametricism. It is a style rooted in digital animation techniques, its latest refinements based on advanced parametric design systems and scripting methods. Developed over the past fifteen years and now claiming hegemony within avant-garde architecture practice, it succeeds Modernism as the next long wave of systematic innovation. Parametricism finally brings to an end the transitional phase of uncertainty engendered by the crisis of Modernism and marked by a series of relatively short-lived architectural episodes that included Postmodernism, Deconstructivism and Minimalism. So pervasive is the application of its techniques that parametricism is now evidenced at all scales from architecture to interior design to large urban design. Indeed, the larger the project, the more pronounced is parametricism's superior capacity to articulate programmatic complexity. [21]

Moreover, parametricism is governed by its own distinctive aesthetic: "Aesthetically, it is the elegance of ordered complexity and the sense of seamless fluidity, akin to natural systems that constitute the hallmark of parametricism."[22]

For the moment, let us overlook the fact that Jean-François Lyotard proclaimed the end of Grand Narratives with the advent of postmodernity—implying that there can be no more universal style any more.[23] Equally, let us overlook the fact that from a technical point of view we need to distinguish very precisely between "advanced parametric design systems" (whose only genuinely parametric manifestations include Catia and Digital Project) and "scripting techniques" (such as Grasshopper), and cannot simply lump them together under the banner of so-called "parametricism."[24] Let us also overlook the fact that most buildings are actually designed these days using explicit modeling techniques, rather than algorithmic generative techniques or parametric tools. Likewise, let us overlook the fact that the notion of style itself seems to remain distinctly postmodern in the way that it privileges the scenographic and the visual.[25] Let us also overlook the fact that the example of Frei Otto, whom Schumacher cites in his article, would seem to suggest that we should be operating in a morphogenetic fashion and using digital tools to perhaps "breed" buildings, rather than simply fashion them stylistically.[26] Finally, let us overlook the fact that much of the work produced under the banner of so-called "parametricism" was already nascent in the pre-computational work coming out of certain offices, such as Gehry and Partners and Zaha Hadid Architects. Let us overlook all these issues, not because they are unimportant, but rather because the focus of this section of the chapter is somewhat different.

What I want to question here is not whether there is such style as "parametricism"—whatever that term might mean—but rather whether there is any such thing as "a digital design" or "a digital building" that might be recognized in terms of its aesthetics or style.[27]

What is "a digital design" exactly? Firstly, I wish to argue that the adjective, "digital," refers primarily to a noun, "tool," used in the process of design.[28] Secondly, we need to make a distinction between the noun, "a design," and the verb, "to design." The noun, "a design," refers to the end product of the design process, whereas the verb, "to design," refers to the actual process of designing. It therefore follows that, if we agree that the term "digital" relates primarily to tools, it is associated more with the *process*, "designing," than with the end product, "a design." We might therefore refer to "designing digitally" when describing the use of digital tools in the *process* of designing. However, we should be very cautious about referring to "a digital design" as such. A building might therefore have been "designed digitally," but—in and of itself—it cannot be described as being "digital." In other words, while there is clearly a practice of designing that involves the use of digital tools, there is no product as such that might be described as "digital." There is therefore no such thing as "a digital design." There is only a process of "designing digitally."

So what exactly do people mean when they refer to "digital design"—or indeed "parametric design"—as a style? Even if there were such a thing as "digital design," as a recognizable product, what exactly would it look like? In short, is Patrik Schumacher correct in identifying a new aesthetic that owes its origin to the development of new digital tools?

The question is set in context by my own personal experience of working with these tools. For the 2006 Architecture Biennial Beijing, Professor Xu Weiguo and I commissioned what we believed would be the first digitally fabricated building in China—a Computer Numerically Controlled (CNC) milled and laser-cut pavilion designed by Elena Manferdini as the West Coast Pavilion. It took us some time to find the right fabrication facilities in China, although we were informed that Yansong Ma, principal of MAD Architects, had in fact located a CNC machine in South China, being used to reproduce antique furniture.[29] We could surmise that this same CNC machine might have been used to also fabricate an entire Palladian villa, in which case our West Coast Pavilion might not have been the first digitally fabricated building in China. What then would be a digitally fabricated design—Manferdini's West Coast Pavilion or a Palladian villa?

By extension, we might ask, whether there is anything that could be drawn using a computer that could not also be drawn using a pencil. The answer, surely, is "no." Equally we might ask whether there is anything that could be fabricated by a robot that could not also be fabricated by hand. The answer again, surely, is "no." There may be certain design outcomes that are facilitated by the use of certain tools, and that might take an impossibly long time to draft in an analog fashion, but this does not make them digital designs.[30] Equally, there might be certain fabrication outcomes whose realization would benefit from robotic construction methods, but this does not preclude the possibility of actually fabricating them by hand, given a sufficiently high quality of workmanship.[31]

The history of architecture is, of course, full of examples of innovations in technology that have a direct impact on innovations in design. For example, the invention of the simple elevator made possible the high-rise buildings that we see in cities, such as New York, today, and it would be fair to say that whole history of architecture changed as a result of that straightforward invention. The same might be said of the air-conditioning unit and many other technological innovations. Equally, the invention of certain software programs has afforded certain design procedures, such as duplicating, populating, subdividing and so on, that has influenced the development of design practices themselves, and—by extension—the invention of certain fabrication technologies has afforded certain construction methods. But neither of these developments actually *caused* these changes. This becomes abundantly clear when we compare a building, such as Gehry and Partners' Guggenheim

Museum in Bilbao, Spain, designed in a largely pre-computational era with one, such as their Disney Hall in Los Angeles, designed comprehensively using computational tools. The two buildings look remarkably similar in appearance.

Figure 4.4 Gehry and Partners, Guggenheim Museum, Bilbao, 1997. Image available at http://commons.wikimedia.org/wiki/File:Guggenheim_Museum,_Bilbao,_July_2010_(06).JPG under a Creative Commons Attribution-Share Alike 3.0 Unported license.

Figure 4.5 Gehry and Partners, Disney Hall, Los Angeles, 2003. Image released into the public domain by photographer Carol M. Highsmith.

So what, then, is "a digital design"? Would we ever talk about "a pencil design"? Presumably not. The term "a digital design" is therefore seemingly absurd. In short, when we understand the term "digital" in architecture we associate it with a series of techniques and technologies—a series of "tools," in other words. These tools *afford* certain operations, and we thereby come to associate them with a certain hegemonic aesthetic that develops over time. But they do not enforce those operations. Thus, computational tools—in and of themselves—do not engender any particular "style."

Toward a theory of affordances

We can therefore see a direct analogy between digital tools and the forms that they afford and architectural forms and the politics that they afford. Architecture—in and of itself—cannot engender a certain politics. The most that it can do is facilitate or hinder a certain politics of use. So too digital tools cannot engender a certain style. The most that they can do is facilitate or hinder the generation of certain forms in both the design and fabrication processes.

If, however, we are to look for a theory that might explain the potentialities offered by digital tools, we might start by considering a "theory of affordances."[32] The theory of affordances suggests that there is a particular action or set of actions afforded by a tool or an object. Thus a knob might afford pulling—or possibly pushing—while a cord might afford pulling. This is not to say that the tool or object has agency as such. In other words, the tool or object does not have the capacity to actually "invite" or "prevent" certain actions. Rather, it simply "affords" certain operations that it is incumbent on the user to recognize, dependent in part on a set of preexisting associations that have been made with that tool or object. Likewise, that action or set of actions is also dependent upon the capacity of an individual to undertake those actions. Thus, certain actions might not be afforded to small children or to those without the height, strength or agility to perform them. Moreover, certain tools afford certain operations, but do not preclude other operations. For example, we might perhaps affix a nail with a screwdriver—albeit less efficiently—if we do not have a hammer at hand. We might also recognize that it is easier to cut wood with a saw than with a hammer, and that the technique of cutting with a saw affords a limited range of possible operations. Importantly also, the theory of affordances has been applied to human–computer interfaces to refer to the easy discoverability of certain actions. As such, we might be able to identify various operations afforded by digital tools that might thereby become hegemonic.

In this sense, computational tools might thereby be associated with a set of operations that are afforded by the accessibility and ease of use of such tools, so that "searching" becomes increasingly popular in a culture colonized by the logic of search engines, such as Google. We might even recognize certain new patterns of behavior, such as the emerging tendency to embark on any research through the use of search engines, so that with time certain hegemonic practices begin to develop and the search engine becomes the *de facto* starting point for any academic research. Inevitably, this will have an impact on the final form that the research will take.[33] By extension, we can observe various visual effects that are afforded by the accessibility and use of the computer in design, so that these visual effects will become hegemonic, thus engendering a certain recognizable aesthetic that we come to associate with designing digitally. Equally, there are certain visual effects that are *not* afforded by certain computational tools. It is, for example, very difficult to generate a straight line using Processing, just as it was relatively difficult to produce carefully controlled curved lines in the analog world of parallel motion drawing boards.

Figure 4.6 West Coast Pavilion, Beijing, 2006. © Atelier Manferdini. Reproduced with permission from Elena Manferdini.

Figure 4.7 Computerized technologies used to produce conventional architectural components. © Dmitry Kalinovsky.

Conclusion

Direct comparisons can be made between the way in which buildings are regarded as "political" and the way in which certain buildings are described as being "digital." Buildings may be *associated* with a certain politics through their appearance—and indeed they *must* be associated with some kind of politics of use —but those politics are not controlled by the form or appearance of the buildings. Equally, buildings may also be *associated* with a certain aesthetic through the digital tools used in their design and construction, but ultimately their aesthetic expression is not controlled by those tools. At best, we might refer to the ways in which a certain politics of use is *afforded* by the form of a building, just as we might refer to the ways in which a certain aesthetic expression might be *afforded* by the tools used in their design and construction. However—in and of themselves—buildings can be neither political nor digital.

We might therefore conclude that it makes little sense to talk about a "politics of parametricism," as though there could be a certain aesthetic expression connected with a certain politics, in that there is no such thing as political architecture, and no such thing as a digital architecture.

Notes

1. In an architectural competition asking for proposals for the building, one architect, Dorin Stefan, even suggested burying the building in sand as a way of returning the building to the hill that had been erased to make space for it.
2. Georges Bataille, "Architecture," in *Rethinking Architecture: A Reader in Cultural Theory*, ed. Neil Leach (London: Routledge, 1997), p. 21.
3. Le Corbusier, *Towards a New Architecture*, trans. Frederick Etchells (London: Butterworth Architecture, 1989), p. 269. *Architecture ou Révolution* was to be the original title of *Vers Une Architecture*.
4. On this see Catherine Cooke, *Russian Avant-Garde: Theories of Art, Architecture and the City* (London: Academy Editions, 1995), p. 118.
5. Fredric Jameson, "Architecture and the Critique of Ideology," in *Architecture, Criticism, Ideology*, ed. Joan Ockman (Princeton: Princeton Architectural Press, 1985), p. 71.
6. Stanley Fish, *There's No Such Thing as Free Speech—And It's a Good Thing Too* (Oxford: Oxford University Press, 1994).
7. Bernard Newman, *Behind the Berlin Wall* (London: Robert Hale, 1964), pp. 31–42.
8. Manfredo Tafuri, in contrast to other commentators, reads the project not from a political perspective, but strictly in terms of urban planning "aesthetic" objectives. Manfredo Tafuri & Francesco Dal Co, *Modern Architecture* (New York: Abrams, 1979), pp. 322–26.
9. Charles Jencks & Maggie Valentine, "The Architecture of Democracy: The Hidden Tradition," in *Architectural Design*, Profile 69 (London: Academy Editions, 1987), pp. 8–25. For Vincent Scully on architecture and democracy, see Vincent Scully, *Modern Architecture: The Architecture of Democracy* (New York: George Braziller, 1974); for Frank Lloyd Wright on the subject, see Frank Lloyd Wright, *An Organic Architecture: The Architecture of Democracy* (London: Lund Humphries, 1939); *When Democracy Builds* (Chicago: University of Chicago Press, 1945).
10. Jencks & Valentine, "The Architecture of Democracy: The Hidden Tradition," p. 25.
11. Fredric Jameson, "Is Space Political?" in ed. Neil Leach, *Rethinking Architecture*, pp. 258–59.
12. Fredric Jameson, "The Constraints of Postmodernism," in ed. Neil Leach, *Rethinking Architecture*, p. 254.
13. Lecture given to the MA in Architecture and Critical Theory, at University of Nottingham, May 1995.
14. Although Bentham's panopticon was never built, the principle of the layout can be seen in numerous buildings, such as James Stirling's Seeley History Library, Cambridge. Here, the control desk is positioned centrally, with all the desks and shelves laid out radially around it, affording an unobstructed view and allowing the librarian to monitor the entire space. A more sophisticated form of panopticism operates with close circuit surveillance cameras.
15. Michel Foucault, *Discipline and Punish*, trans. Alan Sheridan (London: Penguin, 1979), p. 201.

16. Ed. Paul Rabinow, *The Foucault Reader* (London: Penguin, 1991), p. 244.
17. Ibid., p. 245.
18. Ibid., p. 245.
19. Ibid., p. 246.
20. Ibid., p. 247.
21. Patrik Schumacher, "Parametricism: A New Global Style for Architecture and Urbanism," in *Digital Cities AD*, ed. Neil Leach (London: John Wiley & Sons, 2009), p. 15.
22. Ibid., p. 16.
23. On this see Jean-François Lyotard, *The Postmodern Condition*, eds. Geoffrey Bennington & Brian Massumi (Minneapolis: Minnesota University Press, 1988).
24. For a critique of Patrik Schumacher, and an attempt to define precisely the term "parametric" for architectural design, see Neil Leach, "Parametrics Explained," in ed. Kas Oosterhuis, *Next Generation Building* (Delft: TU Delft, 2014).
25. This is not to deny that—as Gilles Deleuze would observe—every process is linked to representation through a process of reciprocal presupposition, and that every process produces a result that constitutes a form of representation.
26. The term "style" implies a preconceived notion of what the end result might look like, as though it constitutes an aesthetic template that constrains and conditions the design process.
27. The term "parametric" seems to cause considerable confusion. It would appear that for many people "parametric design" implies adjusting the "parameters" in the design process. But this makes little sense. From the very beginning of architectural history what building was ever designed without the parameters being adjusted during the design process?
28. Here we might further distinguish between software tools and hardware tools. Although software tools are used primarily in the design process, they may also be used in the fabrication process as in programming robots and so on. By contrast, although hardware tools are used primarily in the fabrication process, they may also be used in the design process in terms of the computers themselves and 3-D printing and other technologies deployed in the refinement of a design.
29. See, for example, the website for alibaba.com http://uk.alibaba.com/product/1777298937-Cnc-architecture.html [accessed June 24, 2014].
30. Let us take the example of a Google "search." We could spend many years searching through all the books in a library to find a particular quote, which we could find in a fraction of a second using Google.
31. This is not to deny that the accuracy afforded by robotic fabrication technologies offers certain advantages over manual construction.
32. The theory of affordances was introduced initially by James Gibson in an article, James Gibson, "The Theory of Affordances," in eds. Robert Shaw & John Bransford, *Perceiving, Acting, and Knowing* (London: Wiley Press, 1977). Gibson later elaborated on this theory in his book, James Gibson, *The Ecological Approach to Visual Perception* (Hove: Psychology Press, 1979). It was also developed by Gibson's wife,

Eleanor Gibson, together with Anne Pick: Eleanor Gibson & Anne Pick, *An Ecological Approach to Perceptual Learning and Development* (New York: Oxford University Press, 2000).

33. By way of example, the translation of Leon Battista Alberti's treatise on architecture was influenced significantly by the then recent introduction of the computer, in that the potential to revise the text again and again (which would not have been so viable in the old days of (re)typing on a typewriter) led to a pithy and precise use of language. Leon Battista Alberti, *On the Art of Building in Ten Books*, trans. Joseph Rykwert, Neil Leach, & Robert Tavernor (Cambridge, MA: MIT Press, 1988).

Chapter 5

Parametricist architecture would be a good idea

Benjamin H. Bratton

In an address to the Council on Foreign Relations on the need for a new geopolitical architecture, the outgoing secretary of state, Hillary Clinton, made a rather striking recommendation. "We need a new architecture for this new world, more Frank Gehry than formal Greek."[1] She described the system dominated by the United Nations, NATO and several other large organizations as the equivalent of the Classical Parthenon in Athens. "By contrast, there's Gehry's Modern architecture. [*sic*] (…) Some of his work at first might appear haphazard, but in fact, it's highly intentional and sophisticated," Clinton said. "Where once a few strong columns could hold up the weight of the world, today we need a dynamic mix of materials and structures." Looking to contemporary architecture for new models for geopolitical "architecture" (literal structures and figurative systems) is—perhaps—a good idea, but it really depends on *what* architecture and *how* it is mobilized toward that end. Regardless of what one may think of Mrs. Clinton or Mr. Gehry's work, why would she offer this kind of architectural analogy at that time? What drives a demand for innovations in the conception of armatures of

---

This text should be taken as a timely intervention meant to influence a wider and still incipient debate on the relationship between synthetic algorithms and the built environment. It is meant to obstruct the motion of this ongoing discourse from one trajectory and toward another, not to articulate the terms of a final outcome. I assume that the reader is familiar with Patrik Schumacher's "Manifesto" text and the most common arguments elaborated on behalf of his version of Parametricism in architecture (global aesthetics, organizational interfaces, a claimed relationship with Niklas Luhmann's sociological systems theory, on the one hand, and Austrian School of economics, on the other, etc.). My own remarks refer to these, and in many cases challenge them, without first repeating them once over.

global power and valuation? The initial emergence of planetary-scale computation as *urinfrastructure* and "information" as a historical category of economic and geographic substance are together the twin engines of the "new world" that Mrs. Clinton is struggling to articulate. But a general and global transformation of hard and soft systems by computation has disturbed this neat arrangement in ways that politicians and political scientists, and designers and design theorists may struggle to understand on its own terms. While State space and borders are perforated and liquefied, State sovereignty and control over information flows is also dramatically reinforced. The possible political geometries at work are twisted and torqued in the extreme.

What does this really look like? The geopolitics of "the Cloud" is an involute geometry made real, but one realized by the overlapping of multiple claims and techniques of many platforms layered one on top of the other. For this, the parliamentary genre of Politics (with a capital "P") as the representation of a sovereign will is diminished by a version of "the political" that is defined instead by rhythmic fields of governance, exclusive neither to States or to Markets, but rather built into ambient interfaces of enforcement, embedded in physical and informational spaces and in their quilting together. In this, Architecture (with a capital "A") may be asked to explore how certain control systems, certain platform systems and specific mereotopological configurations might work toward particular governmental effects. Columns and partitions and/or folds plus punctures multiplied by fields equals what and for whom? At the same time that these architectures are figures reflecting the affective "experience" of the world as it is, or as it may come to be, they are also—very directly—larval schematic variations for both real (if estranged) worlds and their orthogonal futures. They are not symbolic of these, they actually are these. The magnetic polarizations between political and aesthetic registers and between geopolitical form and architectural system continue to wobble and collapse. Well beyond the brief of Clinton's request, these architectures are not only illustrations for organizing geopolitical thought, or drawings on behalf of its potential development. They *are* geopolitical thought in its most direct, compressed expression. If so, is architecture succeeding?

In the name of a speculative geopolitics, my remarks would argue *for* a universal model of parametric design, but one that is only partially recognizable in the one we have now. In short, Schumacher does not go nearly *far enough*. A universalist parametricism would be more radical than what we have seen to date, largely because it would not be, and already is not, limited to architecture per se. That universalist parametricism might be defined as the condition and method of material analysis and composition, endemic to the geologic and geopolitical era of planetary-scale computation, through which algorithmic operations, only partially controlled, can integrate and disintegrate fields of interfaces into platforms, which in turn govern the further distribution of algorithmic events and opportunities. The global transformation of all systems into computational platforms guarantees their potential interoperability and promiscuity, and can also demand or guarantee their quarantine from one another. Exactly *how* these forms of knowledge and technique, the political and the architectural, are being realigned and repositioned and replaced or exploded everywhere by generalized algorithmic logics is perhaps only mappable by algorithmic logics themselves. Their real function may be to index the damage they do. This describes parametric architecture as within a larger field of parametric design, which is itself situated within a larger field of parametric geosystems. It is a pity that this version of parametric design does not exist yet because it is needed for the pressing assignments of geopolitical composition. I mean this in a direct response to how Schumacher delineates his (and its) role, namely that parametricist architecture can be inoculated from the political, which I see as an absurd and self-contradictory suppression of its own ambitions toward universality as a semantics of material communicative systems. At the same time the ultimate expression of parametricist design would not leave our politics intact or be subservient to it. To work, it would reform the political as such, not simply work on behalf of the political-as-is, and so my remarks also are at odds with some of the pointed critiques of Schumacher's positions and of Parametricism in general. Many of these critiques also presume that parametric design and geopolitics are fundamentally separate, but demand that the former has a moral responsibility to be guided by the latter. To be clear, I argue that the geopolitical compositions at work and those most needed are already parametricist, at least according to the expanded and more universalist connotation. To both sides I would argue that neoliberal political economies would be a *casualty* of a fully realized universalist parametricism. Here then, very quickly, are "the why" and "the how" discussed in relation to *interfaces, platforms* and *economics*.

Interfaces

I notice a striking difference between how Schumacher and Teddy Cruz conceptualize the "interface" for architecture. For one, communication in *staged* by architectural interfaces (Cruz); and for the other, architectural space is the *result* of interfacial communication (Schumacher). But either way, space is a carrier of information. We do not need to invoke the "Internet of Things" to see space as informational or interfacial[2] and to be sure a strong suit for parametricist formalism is the rotation of architectural program away from the plan (in essence from the floor) toward a compositional field of interfaces on any or all surfaces at once. I have defined an interface as "any point of connection between two or more systems which governs the conditions of exchange between those systems."³ The interfaciality of any particular point is defined by how it connects or disconnects systems. Some thing is not by itself an interface until it is situated between systems, and so is defined by its interfacial *performance* not by the kind of object that it is. A GUI (graphical user interface) and a doorknob may provide similar kinds of interfacial effects even if they are different kinds of things. And so, within a particular system at a particular time their interfaciality may be, for the designer, interchangeable.

But, despite how Schumacher discusses interfaces as enabling maximal information flow, interfaces do not only grease the path of communication, they also retard it, and that is *good*; it is why you can actually design with them. Some interfaces speed things up *or slow them down*, some are asymmetrical between users and some are symmetrical, some territorialize and some deterritorialize, some make processes opaque and some make them transparent, and so on. Interfaces not only accelerate communication, they also capture it, and so Deleuze's "society of control" is also a theory of urban interfaces. To posit interfaciality only in terms of the stabilization and clarification of space, as Schumacher does, misses the design opportunity. The geopolitical projection of "the architecture" takes place in and through how it activates and deactivates interfaces over time into semi-durable social forms. GUIs, for example, make discontiguous systems legible and instrumental to the user because they are a violent reduction of those systems, and the more violent the reduction, the more instrumental the resulting image-tool. "Click here." Signification and significance; syntax and semantics, this is what interfaces do: interfaces generate potential as much by the modulation of communication not by its simple amplification. This is also one lesson of Koolhaas's "Berlin Wall as Architecture," now played out again all over the world, including in the bifurcation of the San Diego/Tijuana metropolis, the private charter cities and their exclusive walls, and Special Economic Zones along logistics routes. Interfaces at architectural scale allow for social separation between the same population, and *this is their parametric performance and their political effect*. This is not a failure, it is design designating.

This has everything to do with the spatial career of algorithms. I argued many years ago that we should consider architectural design as a subgenre of interface design, and how the generative programmatic doubling of spatial interfaces that mobile media (phones, tablets, etc.) makes into "the new normal."⁴ Forms of spatial organization that society used to ask of architecture, we now ask of software, and so architecture becomes just one particular scale of cloud hardware.

When parametricists talk about the possibility of two or three interpenetrating grids occupying the same space, transforming architecture's own performance as those grids change ratio, we should recognize that this is now the normal condition of urban space everywhere, not a novelty that expert designers might someday bring. What Schumacher calls the "communicative layers" of any urban space are defined by these double-exposed superimpositions of the molecular and the informational as co-promiscuous programs, and the designable terms of that promiscuity is not and should not be contained by the architectural "system." Further, the contiguity of that very system is unwound by its own algorithmization because the shared algorithmic substrate of cloud platforms and cloud hardware—including *architecture*—provides for their composable interoperability, and that interoperability in turn demands their nonexclusivity and the disturbance of their *sui generis* privileges. We should teach designers to think and work this way, to stage interfaces and design according to how they are set in motion and what they do—their dispositions perhaps—and not according to what kind of "thing" they appear to be at first glance.[5]

Platforms

With the regimes of planetary-scale computation, interfaces (in plural) congeal into a higher-order assemblage called *platforms*. Parametricism should also be understood as a design logic for composing platforms and for how particular elements participate in them, but this too demands an alternative conception of the political and the architectonic, and a better and less sentimental language for the geopolitics of information. For example, the conventional State vs. Market distinction is itself an unnecessary and recidivist dichotomy. A strong universalist parametricism would work for and through platforms as an alternative institutional substrate, one that it would also help to realize. Platforms are often a mix of private and public infrastructures, and by strongly enforcing standardized terms of interaction provide bounded exchange between self-directed agents within its network. John Maynard Keynes, Friedrich Hayek, Vint Cerf and Nikita Khrushchev all contribute to the paternity of big platforms, and it is their eccentricity and exteriority from both normative State and Market institutional models—combining elements of these, but drawing as much on basic principles of machine engineering—that has made them so successful in redrawing the effective terms of global architectures, and so also interesting for us (and perhaps Mrs. Clinton) as a general model. Part of their alterity to normal public and private operations is the paradoxical quality of how the centralization and consolidation of platforms can result in massive decentralization of interfaces and users spread throughout the world (and for this, platforms resemble markets), while at the same the coordination of that decentralization reinforces that centripetal centralization of flows, capitalization and exchange into tightly governed platforms (and for this, platforms resemble States).

Let me be specific if not also a bit technical. A working definition of a platform may include references to *a standards-based technical-economic system that simultaneously decentralizes interfaces through their remote protocological coordination and that centralizes their integrated control through that same coordination*. What we call "platform logics" refer to their abstracted systems logic: the diagrammatics of platforms, their economics, their geography and their epistemology of transaction, as well as the tendency on the part of some systems and social processes to transform themselves according to the needs of the platforms that might serve and support them, both in advance of their participation with that platform and as a result of that participation.

In my forthcoming book, *The Stack: On Software and Sovereignty*, I show how some platforms are structured through pyramidal interoperable layers, hard and soft, global and local. Their properties are generic, extensible and plastic, and they provide modular recombinancy, but only within the bounded set of its synthetic planes. It is an auto-generative parametric topography, but one that grows precisely through the autocratic consolidation and rationalization of interfaces and grammars. As suggested, unlike States, that consolidation is driven not from centrally planned legal prescription, but rather through the algorithmic conduction of self-directed behaviors by free-range users. For example, contemporary *Cloud platforms*, such as Google, are derived from specific systems of user-facing interfaces and interactions, and some information "commons" (perhaps preexisting the platform) that provides core content, and that is aggregated, optimized and made more visible, more immediate, more standardized, more interoperable, more mobile and therefore more valuable—to both users and to the platform—than it would be otherwise. As for any infrastructure, it is the simple rigidity of such systems that guarantees access and delimits their platform sovereignties, and it is from the accumulated *accidents* of that accessibility that those sovereignties can be designed. For example, how platforms absorb and recognize patterns in end-user behavior might mimic how markets resolve fluctuations of price, but its formal centralization might also allow for higher-level forms of planning, investment and equity that a State—itself a platform of platforms—is supposed to provide.

Others have suggested that the capacity for a platform to operate in this way suggests some striking similarities with earlier hopes of socialist cybernetics, including Krushchev-era Soviet economists and computer scientists, but also Stafford Beer's *Cybersin* project, commissioned by Allende to design a decentralized cybernetic platform to organize socialist Chile as a vast computational network. The plan was to engineer a pricing and planning mechanism that could observe, analyze, calculate, produce and distribute materials and goods according to principles of rational valuation instead of the anarchic vagaries of supply and demand. Francis Spufford's historical novel, *Red Plenty*, did much to re-spark interest in this overlooked period in the history of political computer science. Back then, planners/programmers had access to what is by today's standards minuscule computing capacity to calculate patterns, pathways and contingencies, but our contemporary supercomputing systems not only can orchestrate and optimize the pricing and dissemination requirements of large economies, they do it every day. (An old joke goes that the greatest accomplishment of communism was the *Apollo* space program. In fact, it may be Google.) In trying to imagine where platform geo-economies can be made to go, and what alternatives to anthropocenic capitalism are possible, it is not suggested that we look back in fondness on mid-century authoritarian regimes for all the clues, but the genealogy demonstrates the point that there is *no exclusive or necessary bond* between capitalism and computational macro-platforms.

This contradicts both Schumacher's claim that there is and must be an exclusive bond between them, which for him is to recommend parametricism, as well as his critics' parallel agreement that this bond is real so as to critique and dismiss it. To a significant extent, the future evolution of multiple and incongruent platforms will work through the transformation of economics by their own technical means, neither State nor market, because platform is State *and* market. Here, the ubiquitous state and the minimal state start to look weirdly similar. Universalist parametricism should be articulated in relation to such platforms. A universalist parametricism does not really want a total deregulation of urban space, despite Schumacher's claims.[6] Rather, it wants platformization, a machine that creates no content of its own but that in the central coordination of mobile agents allows for the rationalization of indexing, forming, folding, enveloping and distribution of energy and matter at various scales. For communism that was once the State, for libertarians it is the market, and *for a universal parametricism it is the platform*. As for geopolitics, this conjunction with platforms does not excuse parametricism from architectural instantiations of sovereignty, on the contrary it defines and demands it.

Toward a universal parametricism?

Perhaps instead of seeing parametricism as nested within a global capitalism, we should see the various species of capitalism as nested within a more general context of planetary-scale computation and generic algorithmic agency. As such, planetary-scale computation would be seen to envelope different genres of capitalization, some of which may or may not take parametricism as a channel for its expression. Any such link is provisional and contingent, and outside that contingency is the space of operations of more universal parametric design, of which parametric architecture is one instance. Policing the boundaries between parametric architecture and parametric design at other scales by tracing the lines by which Modern European politics drew up the division of public institutions and forms of expertise is obviously unhelpful. *Sui Generis* is "nowheresville," it is as a timid response to algorithmic infrastructuralization as fundamentalist-occupying neo-luddism.

Instead, we recognize and leverage other modes of parametric design, such as computational nanotechnology, geoengineering and terraforming, but also the topological configuration of subdivided sovereignties, and locate parametric architecture within this, not adjacent to it nor against it. We have parametric industrial design and interface design obviously. We have parametric logistics (Amazon's warehouses are far more sophisticated spatial expressions of algorithmic design and organization than anything that would work on building envelopes at the expense of the itinerant object). We have parametric politics of the algorithm (demonstrated at the least by State and non-State cyberwarfare, etc.). We have Parametric and algorithmic economics (seen most obviously in the berserker vagaries of high-speed algorithmic trading). We have parametric interfaces (such as Google's individuated maps, which redraw the urban space differently for different users, or better for their own driverless cars). Perhaps most importantly going forward, perhaps more important even than what we call today politics and economics, may be parametric synthetic intelligence, and what we today call robotics. Parametric architecture often already models the user as an automaton; why not go all the way? Why not an urbanism for all the nonhuman users of cities; of which there are always plenty and will be much more? I for one welcome our new robot overlords. Or, perhaps, in the long run anthropometric robotics may prove to be far less transformational than atomic-level machines, what we call nanotechnology (the Visual Arts Department at University of California, San Diego (UCSD) now shares half its space with the Nanoengineering Department, so this topic has come to the fore for me). Considering these convergences with inorganic material, *thanatos* is the wrong diagnosis. It is not a sad destiny to be interwoven with alien matter, and other designed bodies are possible, both anatomic and economic. This human body and this Earthly landscape of matter are only the default settings; they are not destiny, and in the end *chemistry*—already parametric design if there ever was such a thing—may drive the most radical forms of the political imaginary.[7]

In their Accelerationist Manifesto, Alex Williams and Nick Srnicek argue that "Accelerationism is the basic belief technological development can and should be let loose by moving beyond the limitations imposed by capitalist society."[8] Their interpretation is Marxian, in that in the short term, capitalism is what makes these technologies largely possible, but, in the long run, it is what is holding them back from full expression. That full expression not only does not require the continuance of neoliberalism understood as a simple application of dumb markets, but that this logic *cannot possibly survive* the advent of a hegemonic universalist algorithmic design. A universalist parametricism does not need progress as an alibi. It works just as well in relation to anthropocenic catastrophe. We are witnessing tremendous acceleration in computational carrying capacity and at the same time a profound cultural de-acceleration, and these two realities are at least interrelated. An actual for-real-this-time rationalization of built space is the program, but this is only secondarily related to a particular formal-symbolic vocabulary. Versus Schumacher's valorization of parametricism as a High Neoliberal "style" I counter first, that our cities are not nearly degenerate enough, and second, that a global aesthetic for neoliberal architecture already exists and it does not need academic parametricism. Halliburton, McKinsey and Amazon are all also functional parametricists. The market has spoken and it does not care about curves all that much.

I would also caution some of the critics of parametricism that the modern conception of "the Political" will not survive the Anthropocene either. It is useless to erect like a flagpole against what may be seen as gross acquiesce to market conveniences. In the meantime, the political as traditionally conceived is an expression of an institutional division of labor with arbitrary boundaries that are constantly changing. It is always about arbitrary bordering. A strong parametricism, understanding the infrastructural and cultural agency of planetary-scale computation, should be busy problematizing those boundaries, shifting them, perforating them and pixelating them, prototyping them and deforming them, not naturalizing them and reifying them and setting up shop over to one side so as to excuse design from how to designate space. In other words, my own position on parametric design is similar to the apocryphal response that Mahatma Gandhi was supposed to have given when asked what he thought of western civilization, "I think it would be a good idea." Well, I think a parametricist platform logic for spatial interface design based on a strong universalist claims of algorithmic generativity, computational or otherwise, that links architecture to other institutional economies equally transposed by planetary-scale computation, and which sets up the terms for the recomposition of these and their spaces, all toward what I call the post-Anthropocene, *would be* a good idea! One has to wonder why we do not already have that. Why the conservatism, recidivism, irredentism, disciplinary and professional withdrawal and retrenchment? Why the gathering back into a narrower and narrower set of appropriate problems for design to claim?

The intensification of communicative interaction in recent years has largely been an effect of "forms" that exist in realms and at scales to which architecture (as delimited by Schumacher) cannot directly contribute: chip design, software/hardware stacks, wireless network protocols, currency and asset arbitrage algorithms and their effects on land value near trading centers, the microeconomics of supply chain optimization-chasing liquidity, subterranean datacenter design energy-use best practices, relational database optimization, and so on, and on (…). How the further intensification of "communication" (for Luhmann a neutral term, not a goal or ethics) may or may not contribute to what Schumacher calls "the progress of world civilization" at a time when intensive communication is a far-from-suppressed rationale of politics and economics is itself unclear and highly debatable. How any architectonic-geometric form indexes existing communicative flows, optimizes them, disciplines them, organizes them, modulates them, governs them, calibrates them and/or invents new flows remains as important now as ever. It does so, however, precisely because it cannot actually be separated from the effects that it stages (the medium is the message, the syntax begets the grammar, the envelope generates the politics, the mode of production models the social relations, etc.) Architectural form is not *ever* just the "how" of communication. It also always conditions the "what" of communication in any real social context. If this were not so, then the geopolitical and geosociological claims of parametricism as the formal substrate of a rationalized globalization would have no traction and no claim to having any real effect.

A schematic conclusion

As I have argued, Schumacher's limited version does not go far enough in its claim for how parametric and algorithmic logics of form, system, network, etc., can and do model alternative socio-political architectures. A radical parametricism would ultimately explode the legacy distinctions between art, architecture, the urban, politics and law that are held up as Aristotelian ideal realms, and introduce new realms of difference and distinction. The borders of these legacy realms are in fact as plastic and permeable as the membranes of fully parametric forms at architectural scales. For parametricism to make good on its claims, it has to be willing to actually think of itself as a real model for algorithmic geopolitical form in general and in all its guises. Let me say it as clearly as I can: to set the political to one side ("it is not the architect's problem or expertise to deal with the complexities of civil society") and at the same time to make grandiose claims for how architectural form can in fact "remake civilization" is a contradictory and self-defeating point of departure.

I am not arguing for the consolidation or holistic de-differentiation of art, architecture and politics as they exist. I am certainly not arguing for their consolidation in some capital Metaphysical version of these because I do not believe these versions exist (but I do think Schumacher may be turning Luhmann into a Platonist, and at the very least mis-emphasizing the constituted autopoietic form over the constitutive—the emerged over the emergent).

Put most schematically an alternative parametricist manifesto, as yet to be written, might include the following:

1 At this moment, the primary force for systemic formalization is algorithmic emergence, primarily as the transposition of mechanical systems into software systems, and then merging with generalized robotics, and so de-differentiating chemical and informational parametrics.
2 We see this across multiple domains simultaneously, from laboratory science, to warehouse supply chains, to agriculture, to architectural composition, and so on.
3 The formalization of the problems that define these domains into algorithmic logics is supported by the dissemination of media systems based on those logics. Among the most dramatic effects of this is the isomorphic convergence of effects across and between these different domains. Expertise slips and slides from one institution into the other in sometimes odd ways (i.e., the exact same big-data analysis tools and techniques might be applied to urban planning, cellular biology or modeling financial derivatives, etc.). The isomorphic drift of technical-discursive forms is familiar to any reader of Foucault.
4 In this, the formal coherency and diagrammatic borders of differing domains is challenged by the historical transposition of mechanical-labor economies into software-process economies. The original *autopoiesis* of those domains relied on a very different technological substrates, and in some cases, the shift toward software-process economies reinforces old boundaries, and in other cases it perforates them. At many scales new maps, new diagrams, new topologies are required, not just to imagine hypothetical possibilities, but first to make better sense of a reality that

has overgrown those we innovated decades or centuries ago and still keep at our side.

5   In Architecture, parametricism has announced itself as the domain-specific paradigmatic formal language of the general phenomenon of algorithmic emergence. It does so through heterogeneous investigations of formal geometry, figuration, calculation and system, at multiple scales of site and temporality (we hope). We agree that it has important currency in other domains of design as well—urban design, industrial design, interaction design, etc.—but insist that the relative autonomy of these genres of design (as articulated in the early modern era by the differentiation of designer's tools is now reorganized by their shared suite of software tools, many of which are explicitly parametric, and which provide explicitly parametricist designs as their outcome) is less important than what links them together.

6   Architecture's disciplinary and methodological role in this parametric design economy may not be that of a "master discourse" because, as I put it, "all design is interface design" and, as Schumacher put it, "all design is communication design." In this, parametric architecture is only a specific application of the more algorithmic general design logos to architectural problems. As argued above, a universal parametricism is interested in design problems at multiple different registers and domains, especially those that are themselves remade in the image of algorithmic emergence (which is to say most every one).

7   Nevertheless, architecture's expertise with form persists. But in practice, "form is always form *of* something." Except in mathematics—which is not at all the same thing as computation—form is also content. As such the design of form is, in principle, applicable to any socially significant domain that can be organized according to that particular process or gesture, and so the use of parametric logics to compose those forms is not limited to social systems that are themselves already re-determined by parametric logics, but a strong homology can make their compatibility more direct. A universal parametricism would (and actually does already) enter directly into the question of giving form to domains other than architectural forms: GUI and graphics and video games, yes, but also economics (high-speed trading and bitcoin) and politics (TOR anonymity network, National Security Agency data hauls) and many others.

8   Its formal design interventions move into contemporary domains of the political (in the same examples above) that may be almost nothing like the political space of the "parliamentary legislatures, trade unions, and

activist groups" that Schumacher suggests architects go to work with instead of trying to "do political architecture." By comparison with other algorithmic media infrastructure, it is true that TOR and bitcoin, for example, determine how you search or buy, not what you buy, but it would be absurd to say that their "how" does not have political content or program. They work precisely because they are *not* in the proper realm of politics ("Dear Madame Secretary, be it so proposed by this House that there should be a web browser that circumvents surveillance. Let us now vote on it") but they also among many clear examples of a *parametricist geopolitical design*.

9   Put another way, a disinterested artificial intelligence would observe that the most generally efficacious parametric space design software of our moment is Microsoft Excel. This has led globalization to a boring and unsustainable impasse. The impasse—equally aesthetic, political and ecological—demands a more radical response by progressive design. By radical, I mean "to the root," and by "root" we mean the universality of algorithmic logics themselves.

10 It is not that art, architecture, economics, politics, etc., can and should be consolidated into a "Total Plan" (no Lenin and no Le Corbusier). Total plans are antithetical to algorithmic emergence. Parametricism should be more at home with platforms, which are neither State nor Market. Instead, parametricist architecture should be attentive to the need for formal organization within and between nonarchitectural domains that have been disrupted by algorithmic logics both within themselves and between one another (politics and economics included). This is a connotation of Schumacher's definition of parametric architecture as "communication design" that I agree with.

11 The re-differentiations of these domains are not born of the philosophy seminar room but in the world itself, and specifically by algorithmic emergence let loose between them and across them. Philosophy has not caught up to this for the most part. Architectural theory has not either. Economics has led this, but has no idea what it is doing. Politics partially understands, but only on its *para*-margins (TOR, NSA, Google/China, etc.).

12 Parametric architecture is already political whether it wants to be or not, because its actual context as built form is inseparable from its actual context as social machine, and because its formal imaginary can be abstracted to intervene in design problems that organize the political domain directly. You think you are designing an art museum, but your

form may be a new topology for contested air-space jurisdiction over international waters.

13 Yes, the fusion of art, architecture and politics into "one" was the method of mid-twentieth-century totalitarianism, but so was the "fission" of art from architecture from politics. It is the Leni Riefenstahl creative alibi: "Hitler may be the client, but I am just an artist and it is my job to make art and not politics," offering such grotesque cover, and this is still the method of early twenty-first-century totalitarianism ("now is not the time for politics: this is sports, art, entertainment, design, etc.").

14 A withdrawal of parametricism and especially parametricism of form (architecture's expertise) from the design problem of "the political" is both impossible and undesirable. A strong parametricism would not conflate architecture and politics; it would see both of them as algorithmic design problems.

As said, a counter-manifesto on behalf of an expanded parametricism is not written, and perhaps that is because it is not needed. With Schumacher, I share many observations but not necessarily conclusions, and sometimes we share conclusions but not rationales. I enjoy our conversations. As should be clear, I do not believe that the global potential of distributed algorithmic reason as applied to the design and designation of complex systems (including human systems of governance) is well served by any perspective that sees those systems more or less in terms of their early Modern variations and sees computation merely making them "more computational" and more capitalized. At the end of the day, Schumacher's argument may arrive at this stunted conclusion, but if so, then so do the arguments of critics too ready to conflate funny-shaped computer buildings with despotic real estate speculation. If only it were that simple. We can imagine that the algorithmic intelligences reading this book later this century will be confused and even amused by the shortsightedness of some parametric patriots and enemies alike. Fortunately, there are so many other designers and design theorists working to provide them with richer alternative birth stories.

Notes

1. Found at: http://www.ibtimes.com/hillary-clinton-remarks-american-leadership-council-foreign-relations-full-text-1056708 [accessed February 1, 2013]
2. Recall Gregory Bateson's line, "a man a tree and an axe is information."
3. Benjamin H. Bratton, "What Do We Mean by Program? Interactions: Experiences, People, Technology," *The HCI Journal of the Association of Computing Machinery* Vol. XV, No. 3 (May–June 2008), pp. 20–36.

4. Benjamin H. Bratton, "iPhone City, v.2008," in *AD: Digital Cities*, ed. Neil Leach (London: Academy Press, 2009).
5. Siting the "disposition" of active architectural form is perhaps done best by Keller Easterling.
6. Peter Eisenman & Patrik Schumacher, "I Am Trying to Imagine a Radical Free Market Urbanism: A Conversation between Peter Eisenman and Patrik Schumacher," in *Log 28, Stocktaking* (New York: Anyone Corporation, Summer 2013).
7. Reza Negarestani calls this a "culinary materialism," and I have called one method of which this is "rubbing the clinamen raw."
8. Nick Srnicek & Alex Williams, "#Accelerate: Manifesto for an Accelerationist Politics (2013)," in *#Accelerate: The Accelerationist Reader*, eds. Armen Avanessian & Robin Mackay (Falmouth: Urbanomic, 2014), pp. 358–59.

Chapter 6

Play Turtle, Do It Yourself. Flocks, swarms, schools, and the architectural-political imaginary

Manuel Shvartzberg

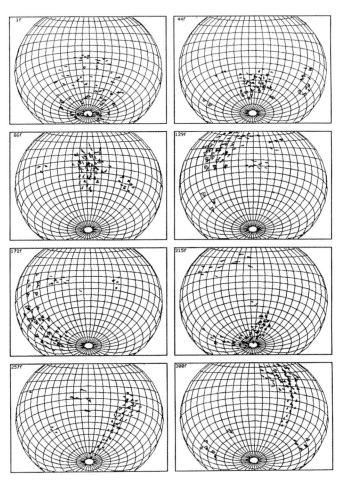

Figure 6.1 Flock simulation by Craig W. Reynolds, 1987. © Craig W. Reynolds.

"It's been found that a few simple rules can regulate very complex behavior. There's a classic computer model for flocking birds, for instance, which only has three rules—keep an equal distance from everyone around you—don't change speed too fast—avoid stationary objects. Those will model the flight of a flock quite nicely."

"A computer flock maybe," Nadia scoffed. "Have you ever seen chimney swifts at dusk?"

After a moment Sax's reply arrived: "No."

"Well, take a look when you get to Earth. Meanwhile we can't be having a constitution that says only 'don't change speed too fast.'"

Art thought this was funny, but Nadia was not amused. In general she had little patience for the minimalist arguments.

"Isn't it the equivalent of letting the metanats run things?" she would say. "Letting might be right?"

"No, no," Mikhail would protest. "That's not what we mean at all!"

"It seems very like what you are saying. And for some it's obviously a kind of cover—a pretend principle that is really about keeping the rules that protect their property and privileges, and letting the rest go to hell."

—Kim Stanley Robinson, *Blue Mars*, 1996

The swarming sublime

In 1987, a historical feat of computer visualization and zootechnological inquiry was accomplished: Craig W. Reynolds, a computer graphics engineer, presented the first working digital simulation of a flock of birds. Reynolds, who graduated in 1978 from the Massachusetts Institute of Technology's (MIT) Architecture Machine Group,[1] showed how a set of simple rules—keep an equal distance from everyone around you; don't change speed too fast; avoid stationary objects—applied to an agent-based modeling program could achieve the apparently synchronized, decentralized, and emergent behavior that we today associate with a common trope of the architectural imaginary: the swarm. The ramifications of this simulation went far beyond its original context—research in computer graphics—to affect not only architecture, but also fields such as Artificial Intelligence (AI), cognitive science, evolutionary biology, the entertainment industry, and sociology.[2] In this chapter, I will explore the historical and theoretical relations between swarm simulations and the political imagination: a relationship that is often rehearsed, consciously and unconsciously, within contemporary architectural discourse—and one of the principal figurations within the discourse of "parametricism."[3]

The roots of Reynolds's achievement, like so much computer science within and without architecture, are to be found at MIT in the 1960s and 1970s. Reynolds programmed his agents through the appropriation of "Turtle Geometry"—a pedagogical and computer programming technique that had first been developed in the late 1960s by Seymour Papert, a mathematician and co-director with Marvin Minsky of MIT's AI Lab. Papert's turtles, however, did not fly; they began as semi-automatic robots equipped with wheels, motors, sensors, and pens that crawled slowly around the floor drawing pictures, performing the seemingly mundane yet highly complex task of, in Papert's words, "teaching children thinking."[4]

The history of digital swarm simulation is therefore tied to ideas of cognitive science, developmental psychology and education, as much as it is to ideas of organicism and computing. All these different realms and practices of knowledge came together at MIT within the context of the US military-industrial-academic complex, charting a tortuous course that ran from the deep engagement with national security projects in the 1970s, to a preoccupation with economic competitiveness in the world markets of the 1980s. As it turns out, schools and swarms are slightly at odds with each other in this story: Papert's revolutionary work with children in the 1970s provides an unwitting, but necessary, techno-ideological and historical backdrop for the development of global capitalism's self-understanding in the 1980s—the ideal and concrete reality of market flows as vital, unregulated swarms: an image and reality well-nigh understood and leveraged by the master-shepherd of the techno-flocks, Nicholas Negroponte, toward the establishment of MIT's Media Lab in 1985. Thus, the politics and aesthetics of research are "imagined"—technically and materially speaking—according to the historical economic and epistemic configurations at play.

How these *images* of swarms are constructed and how they, in turn, feed back into *political imaginaries* is the general subject of this chapter. More specifically, I will explore how Papert's 1970s techno-scientific work with children, infused by a pragmatist educational philosophy, unintentionally generated the conditions of possibility for the image of the swarm to be both technically feasible and ideologically desirable as a cipher for global capital in the 1980s (and beyond). Papert's work drew a techno-political circuit that linked the child's motor skills and subjectivity to ideas of what constitutes legitimate knowledge, creativity, and autonomy, suggesting a biopolitical shift toward a new *para*-metric of labor, thus redefining notions of productivity and practices and techniques of accountability—quite against the grain of the rigid existing educational and institutional structures. In turn, however, this work was key for the production of an image of the (adult) individual *as an organic, self-sufficient unit* within a meta-organic system of capitalist competition, deregulation, privatization, and accumulation in the 1980s—a system that cannibalized the very schools where Papert's original "revolution" had begun.

The swarm, in this history, is the concrete aesthetic and functional operator of a political imaginary based on the idea of "the public" as an indeterminate aggregation of natural, private interests rather than an artificial, social achievement. In this sense, it can "stand in," metonymically, for the global capitalist market. But other than as semiotic trope or systemic operator, the aesthetics of swarms also perform another kind of work. Already in the eighteenth century the philosopher Edmund Burke pointed out the particular kind of *sublime* associated with the aesthetics of configurations based on large numbers. He called it the sublime of "magnificence,"[5] owing to its characteristic interplay between order and disorder, and the dazzling spectacle of profusion and multitude—all attributes that configure the aesthetics of swarms. The magnificent, according to Burke, is a kind of abandonment of the senses in thrall of these attributes. And, it is this abandonment that constitutes the other kind of work that swarms appear to do. This work is not constituted by the system of order and operation—the intrinsic functional dimensions—that drive swarms (or their simulations), but is rather an index of a more metaphysical kind of drive; a drive toward the metaphysical itself. On the one hand, images of swarms provide instances of recognition of a grave power that is, to a certain degree, daunting but pleasurable—such is Burke's definition of the sublime—but on the other hand, they illustrate a power necessary for us to recognize *our place in the cosmos*: an overall sense of direction; a relationship between self and other, individual and group; and at the most fundamental level—an image of "nature."

The images of swarms discussed here powerfully conjoin aesthetic and political dimensions to produce an *imaginary*: they reify the central question of sociology—the relationship between individual and society—which is always constructed in a particular historical-political context, by elevating it to the level of the universal.[6] The repetition and recirculation of images of swarms in architectural (and other) discourses also performs this universalizing role: a profusion of images of swarms that quasi-organically welds the notion of "magnificence" to *an infrastructure of display*, and therefore, to *a global economy of "sense."* This relationship between circulation and meaning, rendered as an organic process, constitutes the most important political dimension of the imaginary of swarms. The endless repetition of representations to the effect that society is a swarming aggregation of individual self-interests is an index of the political power at stake in this imaginary. Faced with this pervasive power, one possible way not to succumb to its calls for abandonment—in other words, to demystify, deflect, or redirect this power—may be to consider how images of swarms are culturally and technically constructed, rather than plainly accepting them as "natural." This means understanding the technical and historical processes by which swarm images—or simulations—emerge and are mediated.

The rise of the boids

Craig W. Reynolds's 1987 flock simulation is a case in point. Using software developed from Seymour Papert's original Turtle Geometry program (called "Logo"), Reynolds invented a "model [that] has often been cited as an *urtext* of computer-assisted *biological* swarm research."[7] As Reynolds stated:

> Geometric flight is the basis of "turtle graphics" in the programming language Logo. Logo was first used as an educational tool to allow children to learn experimentally about geometry, arithmetic, and programming. The Logo *turtle* was originally a little mechanical robot that crawled around on large sheets of paper laid on the classroom floor, drawing graphic figures by dragging a felt tip marker along the paper as it moved. Abstract *turtle geometry* is a system based on the frame of reference of the turtle, an object that unites position and heading. Under program control the Logo turtle could move forward or back from its current position, turn left or right from its current heading, or put the pen up or down on the paper. The turtle geometry has been extended from the plane onto arbitrary manifolds and into 3D space. These "3D turtles" and their paths are exactly equivalent to the boid objects and their flight paths.[8]

Thus Papert's turtles "learned" to fly in swarm formations by being turned into "boids" (bird-like objects) and applying the vector and object-oriented programming methods inaugurated by Logo to configure a set of simple rules for regulating their aggregate motion. Rather than attempting the cumbersome and historically intractable process of modeling the whole swarm, Reynolds's cunning lay in programming each individual boid as a Logo turtle and setting a number of individual constraints in relation to the whole (i.e., all boids should tend toward the geometric center of the flock), thus allowing the computer to iteratively calculate the resultant aggregate of each individual boid's movements.[9] The result was surprising for Reynolds (and the world), as swarm dynamics were successfully replicated on screen, and as the boid flocks engaged in seemingly "unplanned" and intentional individual and collective behaviors, such as maneuvering around obstacles, breaking up in independent groups and then rejoining together, as well as abrupt and sudden changes in direction.[10]

Figure 6.2 Reynolds's boid flock circumventing obstacles. © Craig W. Reynolds.

The simulation thus appeared to demonstrate that, as Reynolds put it, "the aggregate motion that we intuitively recognize as 'flocking' (or schooling or herding) depends upon a limited, localized view of the world,"[11] as the boids programming was geared primarily toward individual rather than collective, global properties. For this reason, Reynolds's simulation became a much-cited example for the phenomena of "emergence" and "self-organization" across a number of disciplines.[12] As a sociological text that cited Reynolds's simulation explained, "the rules do not, in themselves, predict flock behavior; the exact form of the structure, the flock, is neither entirely random nor predictable."[13] Crucial to this definition of "emergence" in complex systems, then, was the impossibility to accurately determine or predict the final outcomes of the model, as these emerged out of states and variables that depended on the iterative evolution of the model itself.[14] Reynolds's simulation showed how a self-organizing (in the sense of autonomous and decentralized), virtual system could in fact approximate, and in some ways imitate, life itself—including living systems' nondeterministic, random, chaotic, and yet, ordered behaviors—and therefore suggested a powerful conceptual tool for modeling other complex social, physical, and organic processes, inspiring new methods for the investigation of so-called "artificial life."[15]

The boids simulation thus provided fodder for the ever-elusive attempts to reproduce "intelligent life" by artificial means; a marginal but persistent thread of the computational imaginary at least since Alan Turing's question, "Can machines think?"[16] Yet, while Turing's landmark 1950 article astutely displaced questions of truth, authenticity, and hermeneutics in favor of a purely performative and operational definition of "thinking," this was not always the case for AI enthusiasts.[17] Reynolds's self-organizing aggregate of boids, for instance, opened up the quasi-magical possibility of reverse-engineering the "intelligent" automaton by way of the disaggregation of the boid flock. A reverse-engineering that would deduce the free-willing being (boid or other) from the aggregate organic whole via a metaphysics that identified "the individual" and "the collective" as organic poles of the same "natural" system. If only such reverse-engineering were possible, Reynolds's simulation seemed to suggest, AI might still be capable of creating autonomous intelligent artificial beings.

Interestingly, though, the opposite is rather more historically accurate. The boids did not have "a mind of their own"[18] (as Reynolds quipped) due to the emergent properties of their programming (not exclusively, anyway), but were in fact themselves modeled on the idea of people as independent and autonomous beings that could be "aggregated" like meta-biological organisms.

Not any kind of people either, but a particularly susceptible kind for defining and creating new forms of life—children.

Play Turtle, Do It Yourself

First developed by Seymour Papert at MIT's AI Lab in the late 1960s, the Logo program established the conditions of possibility for the 1987 computerized swarm. Papert, then a young South African mathematician with a strong interest in developmental psychology, arrived at MIT after spending the years between 1958 and 1964 working with Jean Piaget at the University of Geneva. At MIT, Papert joined the Research Laboratory of Electronics' Project MAC—for Multiple Access Computer or Machine-Aided Cognition—out of which emerged the AI Lab, originally founded by Marvin Minsky and John McCarthy. Project MAC was heavily funded by the Advanced Research Projects Agency (ARPA, precursor to DARPA), set up by President Eisenhower in 1958 to invest in military-industrial scientific research.

Papert worked with Daniel Bobrow, Wally Feurzeig, and Cynthia Solomon at the Cambridge-based high-technology firm Bolt, Beranek & Newman and the AI Lab to develop Logo, a Lisp programming language. In this context, Papert and his colleagues sought to elaborate their own educational program, in both computational and pedagogical terms. Drawing from MIT's vast melting pot of electrical engineers, linguists, information theorists, behavioral psychologists, mathematicians, and others, Papert began to formulate a project that would connect some of the ideas and techniques being developed on cognitive science, computing and cybernetics, to experiments with child learning. Logo was born, therefore, from the intersection of quite different fields of culture and scientific study—from the military-industrial-sponsored research on computing and cognition, to Piaget's theories of developmental psychology, to the general countercultural, anti-institutional atmosphere of the 1960s at MIT.

In Logo, these different tendencies and forms of knowledge all purposefully crystallized as one project. Attempting to reframe the dominant contemporary notions of child learning, Papert used the work on cognition, apprehension, and cybernetics being developed at the AI Lab to advance a more "holistic" and philosophically "pragmatist" approach—influenced by John Dewey's ideas of learning through experience—to learning mathematics. As he explained in a 1973 funding proposal for the National Science Foundation, the "essential educational content," he was interested in developing, lay in "'re-conceptualizing' areas of knowledge such as physics, mathematics, music, physical skills, elementary cognitive science, elementary computer science (including programming)"[19] rather than merely prescribing a new mathematical curriculum.[20] "The flavor of turtle geometry," Papert noted in relation to traditional mathematics teaching methods, "can be tasted by comparing typical descriptions of a geometric figure, say a circle, in Turtle Geometry and in Cartesian Geometry:

Cartesian: $y - b + \sqrt{R^2 - (x - a)^2}$

Turtle:
TO CIRCLE
FORWARD 1   This is a program
RIGHT 1   written in Logo to
CIRCLE   generate a particular
END   sized circle."[21]

This difference in method was not simply a formal one, but was the very substance of the epistemological shift Papert was arguing for. "The cognitive theory underlying our work," he argued, "draws on ideas from the Piagetian tradition of thinking about children and from these aspects of Artificial Intelligence concerned with thinking about thinking in general."[22] This approach drew heavily from computer programming languages, combined with robotics, as a method for getting children "to *do* mathematics rather than merely to learn *about it*."[23]

Papert sought to implement this pedagogical-cognitive experiment by teaching children to program their own computer graphics, thus getting them to learn geometry "by doing," rather than by "drill and practice,"[24] an approach greatly indebted to his experience with Piaget at the International Center for Genetic Epistemology in Geneva. Piaget proposed a "stagist" conception of the child's development—certain cognitive functions, such as abstract thought, could only occur after a certain age; a conception that Papert criticized—but one of his main ideas was that the child learns by assembling her own epistemological frameworks according to what is around her. As Papert noted, he "came away impressed by [Piaget's] way of looking at children as the active builders of their own intellectual structures,"[25] prompting Papert to develop this idea by focusing on the cultural-material context that the child uses to "construct" her frameworks of understanding—an idea he later thematized as "Constructionism" to define his own techno-pedagogical theory.[26] "For example," Papert wrote, "the fact that so many important things (knives and forks, mothers and fathers, shoes and socks) come in pairs is a 'material' for the construction of an intuitive sense of number."[27] Thus, Papert elaborated,

"Piaget's work on genetic epistemology teaches us that from the first days of life a child is engaged in an enterprise of extracting mathematical knowledge from the intersection of body with environment."[28] The question, therefore, was how to get the child to become conscious of the fact that she was already formalizing mathematical knowledge, albeit subconsciously.

To do so, the Logo group postulated that getting children to program a computer would subtly engage them in systematic and meta-cognitive thinking: "teaching the computer how to think, children embark on an exploration about how they themselves think. The experience can be heady: Thinking about thinking turns the child into an epistemologist, an experience not even shared by most adults."[29] Papert's computational-pedagogical project was thus formed around three conceptual nodes for encouraging self-conscious systematic thought, the terminology of which he borrowed from Freud as well as Piaget: Logo was conceived as a vehicle for "the child to become personally—intellectually and emotionally—involved"[30] in her own learning experience. The objective was to combine the Piagetian spatial-mathematical knowledge the child had already subconsciously acquired—which he termed the "body-syntonic"—with the idea that allowing children to express their own interests and enthusiasms would lead to more motivated, responsible, creative, and systematic behavior—what he termed, following Freud, the "ego-syntonic."[31] And, in between these two registers, the purely geometric (external) and the purely psychic (internal), Papert conceived of a mediating techno-social interface that he termed "cultural syntonicity"—and which was modulated by a "peripheral" in the form of a cybernetic, robotic turtle:[32]

> In Logo environments many children have been started on the road to Turtle geometry by introducing them to a mechanical turtle, a cybernetic robot, that will carry out these commands when they are typed on a typewriter keyboard. This "floor Turtle" has wheels, a dome shape, and a pen so that it can draw a line as it moves. But its essential properties—position, heading, and ability to obey TURTLE TALK commands—are the ones that matter for doing geometry. (…) Since learning to control the Turtle is like learning to speak a language it mobilizes the child's expertise and pleasure in speaking. Since it is like being in command, it mobilizes the child's expertise and pleasure in commanding. To make the Turtle trace a square you walk in a square yourself and describe what you are doing in TURTLE TALK. And so, working with the Turtle mobilizes the child's expertise and pleasure in motion. It draws on the child's well-established knowledge of "body-geometry" as a starting point for the development of bridges into formal geometry.[33]

Figure 6.3 Seymour Papert with robotic Logo Turtle, c. 1970. Photograph by Susan Pogany, Courtesy MIT Museum.

This techno-cognitive apparatus sought to project, for the child, an "image of himself as an intellectual agent."[34] From 1967, the Logo team began deploying Turtle and Logo experiments in custom-designed courses for groups of children at different schools around Massachusetts, and from the 1970s and 1980s, around the country and the world. Each course was an opportunity to try out different iterations of turtle robots, Logo program versions and also, pedagogical tactics, exercises, and techniques.[35] These teaching experiences informed the AI Lab's development of the project, in both technical and pedagogical terms, and were also the basis for the multiple National Science Foundation funding applications the group successfully applied to in the 1970s.[36]

The hands-on teaching experiences in schools appeared to prove the efficacy of the "constructionist" techno-pedagogical theory. Using the turtle to draw shapes, Papert argued, the child is "teaching the computer," which "objectivizes the [learning] process."[37] Mobilizing the combination of individual movement (the "body-syntonic") and volition (the "ego-syntonic"), Logo enabled the drawing of geometric shapes "created in reference to the student's own body instead of in reference to a Cartesian coordinate system, or an extrinsic point of view"[38]—interpellating the child as an autonomous subject standing at the center of its own world. In other words, this way of learning encouraged the child to project herself as a radically "free" and "desiring" subject. Or, following Papert's succinct slogan: "*Play Turtle. Do It Yourself.*"[39]

Perhaps the most significant work of Papert's Logo turtle system happened here, when the child discovered her own "pleasure in commanding." Despite the potentially authoritarian connotations of these statements, however, this was not an ideological project of indoctrination—it was, strictly speaking, a technological apparatus: "The 'Logo Turtle Lab' is a time-shared computer facility (a PDP-11/45) equipped with an unusual variety of 'peripheral' devices: graphic devices (CRT displays, plotter), sound making devices ('music box,' phoneme generator) and mechanical devices ('turtles,' motors)"[40] with which the child simply "plays Turtle."

In this sense Papert's turtles represented a radical pedagogy indeed: his program had the effect of getting children to produce—and to radically exercise—their own "free will" as they commanded the turtles, rather than getting them to merely repeat things or learn pre-established procedures in the manner of the conventional mathematics curriculum. This, then, ultimately constitutes Logo's meta-cognitive dimension that folds the pedagogical project back into the broader framework of the AI Lab's search for autonomous AI. Playing Turtle, the child not only builds her own mathematical-geometric formulations, she also performatively "learns" a particular mode of individuation and socialization predicated on creatively teaching herself and informally teaching, commanding, or collaborating with others—a techno-pedagogy that is suitably reproduced through the recursive processes of individual desire and computer programming.[41] The turtle engages the projected ego and motor skills of the child, "thereby gaining a greater and more articulate mastery of the world, a sense of the power of applied knowledge and a self-confidently realistic image of himself as an intellectual agent."[42]

The Logo system thus constitutes a veritable apparatus of subjectivization: plugged in to the time-sharing computer and its consoles, stimulated by the optical signals on the screen and the drawing paper on the floor, and free to "command" the cybernetic turtle at the tip of the whole ensemble, the child learns to act as a self-directed agent; seeking his own technical means by which to satisfy his desire; as an "entrepreneur of himself," to use Michel Foucault's apt formulation.[43] In this setup, the child, via the necessary turtle, becomes an abyss of intentionality—a source of self-perpetuating procedurality and meaning arising from self-satisfaction: the child aspires to do what he wants to do which is what the apparatus enables him to aspire—a recursive statement in the form of the turtle-program-child assemblage that is literally "realized" through the "emergent" performativity of the drawing-programming process.[44] Papert's transcriptions of these processes into statements and tabulations of value close the circle by bureaucratizing it, inaugurating the era of "measurable"—and thus accountable—techno-creativity.

Logo's numerical-aesthetic enactments—the desiring self's "magnificent" sublime recursions[45]—led Papert to theorize the potential for eventually multiplying, clustering, and animating groups of virtual autonomous turtles themselves—and thus anticipated Reynolds's simulation up to fifteen years in advance.[46]

The Media Lab: Aggregation via disaggregation

As Papert speculated in the 1973 Logo Memo, "Uses of Technology to Enhance Education," the idea of a "Build-An-Animal Kit" with cybernetic turtles could be taken a step further in abstraction to become the theoretical template for modeling computer-simulated, aggregate biological processes. By the mid-1980s, while Papert's research continued to focus on the development of Constructionism, this virtual "aggregation of animals"[47] idea took shape as an ambitious new project headed by Marvin Minsky and Alan Kay within the new Media Lab. Called "Vivarium," the project brought together students from different disciplines with a singular purpose:

> Mission: create "life." Enable school kids to invent and then unleash realistic organisms in whole "living" computerized ecologies—learn about the universe's creation by doing some of their own. The animals they create would behave, learn, even evolve independently.[48]

The Vivarium drew its students, skills, and resources for aggregating "whole 'living' computerized ecologies" from another aggregation process: the Media Lab's agglomeration of disciplines under the single roof of the new Wiesner Building, designed by I.M. Pei & Partners and opened in 1985. A veritable brainchild of Nicholas Negroponte, the Wiesner Building represented the culmination of a general dynamic of disciplinary convergence at MIT in which the AI Lab, the School of Architecture and Planning, the Center for Advanced Visual Studies, and other departments had been involved with for over fifteen years—not without institutional strife.[49] Negroponte's own Architecture Machine Group ("Arc Mac," established in 1968) was an early force in this integrative drive, having been conceived through Negroponte's collaborations with the Department of Electrical Engineering, the AI Lab, and the School of Architecture. This integrative drive, however, was not just the result of MIT's academic institutional-disciplinary dynamics, but reflected broader dynamics of national and international economic, political, and also aesthetic reconfigurations.

By 1973, MIT's leadership had begun to consider ways to integrate the arts, technology, and sciences across campus—an idea that crystallized when philanthropists Abe and Vera List committed $3 million for a new arts gallery in 1977.[50] Negroponte, however, independently approached MIT president Jerry Wiesner with a different plan to develop a more "entrepreneurial" program combining MITs strengths in science, media arts, design, and technology. "Surely," Negroponte speculated, "we at MIT could do much better [than an arts gallery], something at a more Olympic level, something that was at the cutting edge of both art and technology, something that would be less like chamber music and more like science fiction."[51] The reference to Olympic and science fiction ambitions were not casual—they were designed to mythologize Negroponte's own research. Arc Mac's experimental projects with computer interfaces and its exchanges with the AI Lab had established Negroponte as a key figure in the shifting institutional landscape of MIT in the 1970s, which he leveraged "toward a theory of architecture machines" in which such machines might one day learn, self-improve, be ethical, and "appreciate the gesture."[52] These alluring ambitions, together with his academic and reputational links, including his close ties to the military-industrial-research complex and the philanthropic apparatus of MIT—which Negroponte ably mastered, as his close association and subsequent friendship with powerful figures like Wiesner would prove—were decisive in his securing the leadership position to establish the aggregation machine that would become the Media Lab.

As Marvin Minsky noted regarding the shifting funding patterns for experimental research in the United States, MIT's early postwar ties to the federal defense establishment had provided the AI Lab and others (including Negroponte) an almost informal and patron-like relationship to easy funding[53]—a state of affairs characterized by historian Paul Edwards as a "closed world," due to the close-knit and elitist allocation of government's research capital through highly discretionary networks of association.[54] In this context, Arc Mac's projects, all heavily funded through federal military research (primarily via DARPA, the Office for Naval Research, and, to a lesser degree, the National Science Foundation[55]), ought to be read in light of the increasing security anxieties posed by both national and international urban-political struggles of the 1960s and 1970s.[56]

Addressing these concerns, Negroponte's experimental projects (such as SEEK, 1970) leveraged the aesthetics and technics of agent-based aggregations to theorize a machine-enabled "participatory design" that would assist or take over the entire design process as a way of channeling and containing urban discontent.[57] As Negroponte argued, by effectively monitoring and coding the user's individual behavioral parameters into a comprehensive and systematic computational-robotic program, "the design of the city can start to reflect every single inhabitant—his needs and desires," with no need for "middlemen" such as architects or politicians.[58] The aggregation of individuals' needs and desires, "objectively" measured and interpreted through empirical quantitative and systems data analysis, would render the city's environment transparent—to the individuals themselves, but also and especially, for the benefit of the environment's adequate self-governability.

Negroponte's "participatory design" would therefore be of great interest to the Department of Defense as it suggested the possible "stabilization" of urban environments upended by riots and insurrections, whether in Vietnam or the US, via a "self-organizing" process that intrinsically upheld acceptable systemic parameters, and thus invisibilized apparent external influences.

However, as perceptions of security threats in the country changed, so did the politics of funding and its projects. By the early 1980s, a marked shift in the sources of American insecurity and anxiety had taken place, turning from a paradigm of national and international "stabilization" fought by the military-industrial-research complex, to one in which the United States was perceived to be at risk of failing to compete within and through the increasingly deregulated international economic markets.

From the urban and territorial volatility experienced in the 1960s and early 1970s—symptomatically addressed by MIT's various projects of participatory environmental control and management—the turn of the decade appeared to signal a new field of struggle in which direct military power and intelligence—and

Figure 6.4 The Architecture Machine Group's SEEK, at the "Software" exhibition, 1970. Image provided by the Jewish Museum, New York.

thus, DARPA's research money—had lost its former effectiveness. The new weapons and terrain of hegemony, rather, were the advanced skills required for workers to be able to compete in a now truly global market economy—in other words, the broad terrain of education. Haunted by the rapid rise of Japanese industrial power and an apparent abundance of evidence showing a long-drawn and persistent dynamic of underperformance in US schools, the national anxiety agenda came to be dominated by the question of how to reform the schooling system to equip students with the right skills for competing in the new "information age" economy. In this context, President Ronald Reagan commissioned a report from a specially created National Commission on Excellence in Education to review the situation. Called *A Nation at Risk*, the report was published in 1983 and became a highly influential, strategic rhetorical weapon within the ensuing political struggles over the nature and funding of education, providing substantial ballast for neo-conservative reform attempts of the schooling system, for the sake of the United States' world hegemony. As the first page of the report starkly asserted:

> Our Nation is at risk. Our once unchallenged preeminence in commerce, industry, science, and technological innovation is being overtaken by competitors throughout the world. (…) What was unimaginable a generation ago has begun to occur—others are matching and surpassing our educational attainments. If an unfriendly foreign power had attempted to impose on America the mediocre educational performance that exists today, we might well have viewed it as an act of war. (…) We have, in effect, been committing an act of unthinking, unilateral educational disarmament.[59]

Such "disarmament" thus called for urgent measures. However, rather than focus on addressing the problem through more comprehensive school funding, the report subtly—and not so subtly—displaced the schooling system's inadequacies, such as those related to the subjacent racial and economic exclusions that had inflamed the country's cities over the past two decades, to the abstract and de-historicized issue of "accountability" (testing and performance criteria). In doing so, the report tacitly blamed the desegregation efforts begun with the civil rights movement for the "slump" in academic achievement: "*the average graduate* of our schools and colleges today is not as well-educated as the average graduate of 25 or 35 years ago, when a much smaller proportion of our population completed high school and college. The negative impact of this fact like-wise cannot be overstated."[60]

Accordingly, as Reagan took power he argued that it was the federal education programs implemented in the wake of President Johnson's "War on Poverty"—such as the landmark 1965 Elementary and Secondary Education Act—that had caused the performance problem in the first place, and thus called for the total elimination of the US Department of Education. Upon taking office, his administration implemented sweeping reforms of decentralization, deregulation, defunding, and privatization of the school system. Thus, the oppressions and exclusions of the 1960s and 1970s, the roots of the "urban crisis" that had swept the nation, were displaced onto the terrain of education and employment: as schooling became more and more a target of the neo-conservative movement, a parallel and related move deracinated industrial cities through the international outsourcing of jobs where cheaper labor could be exploited.

These specific school policies did not affect MIT directly—but the Institute played a tacit role in the country's drive toward the implementation of a new mode of production based on the "information age," generally oblivious to its political dimensions. In this context, the Media Lab's new funding strategy was symptomatic of the shift—not only illustrating, but performing it.

By the late 1970s and into the 1980s, DARPA's funding toward MIT's most experimental research had dwindled. Negroponte, however, already deeply involved in fundraising efforts toward the new Media Lab, responded to this challenge by turning to private industry. This was both a necessary condition imposed by the wider cuts in public funding and a condition imposed by the MIT Corporation "to raise money from 'new friends,' not tapping into the traditional channels of MIT endowment,"—as Negroponte acknowledged, "Philanthropy is very much a zero-sum game."[61] The shift toward seeking funds from the private sector, however, also reflected a new reality for industry: the convergence of media, industrial, and computing companies into large conglomerates with a vital interest in expanding their businesses.

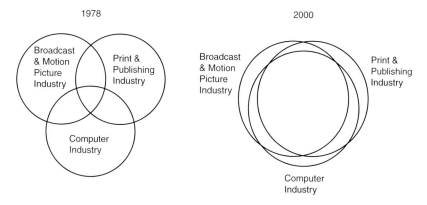

Figure 6.5 Nicholas Negroponte's diagram of media technologies convergence, c. 1978.

The future Media Lab thus presented itself as a perfect innovation "partner" for a media-business landscape in the cusp of momentous corporate shifts, a period of new media and technology agglomerations, bankruptcies, and intense competition—as described by MIT's Political Science Chair, Ithiel de Sola Pool, in his 1983 book on media convergence, *Technologies of Freedom*.[62]

Thus, as public funding disaggregated, Negroponte turned to a funding model of competitive, private aggregation, pitting important media companies against each other for "access" to the Media Lab's innovation riches. The Wiesner building was a product of aggregation through disaggregation: pivoting from his previous federal research-funding connections, "new friends were found" by Negroponte "in three places: Hollywood, Japan, and the personal computer industry."[63] These "new friends" formed a huge international network of industry and government, the singular pieces of a new, swarm-like geopolitical economy:

> Nobody political has paid the slightest bit of attention to Negroponte or the Media Lab. Who's lined up at the door is the business studs. General Motors, ABC, NBC, CBS, PBS, Home Box Office, RCA, 3M, Tektronic, NHK (Japan's public TV network), Ampex, Harris, Mead, the Defense Advanced Research Projects Agency (DARPA), IBM, Apple Computer, Warner Brothers, 20th Century–Fox, Paramount, LEGO, Dow Jones, Time Inc., Polaroid, Kodak, Schlumberger, Hewlett-Packard, Digital Equipment Corporation, BBN (Bolt, Beranek & Newman), The Washington Post, The Boston Globe, Asahi Shimbun, NEC, Sony, Hitachi, NTT (Japan's AT&T), Sanyo, Fujitsu, Fukutake, Bandai (Japan's largest toy maker), Mitsubishi, Matsuhita—a hundred sponsors, a few of them government, most of them corporate.[64]

As a senior IBM manager who dealt with Negroponte highlighted, the latter possessed "real world-class salesmanship," clearly relishing the opportunity to mingle with the (also "closed") world of "business studs" in high corporate and philanthropic power.[65] From elitist origins himself (his father was a Greek shipping magnate), Negroponte was at ease among the people he dealt with through fundraising; meeting "extraordinary people, almost all of whom were friends of or contacts through Jerry [Wiesner]. It is obvious when you say it: people in a position to give serious money are leaders, achievers, and very interesting people."[66]

The deal offered by the Media Lab to prospective corporate sponsors was predicated on the perceived competitive advantage of *access* to "blue-sky" experimental research that corporations could not usually afford the risk to do themselves. While all the research was nonproprietary—meaning that it could not be exclusively dedicated for commercialization by any one corporate funder—the sponsors benefitted from having "a five-year key to the lab" in which they could freely "wander around and ask questions about the different projects."[67] In this gamble of selling mere "access," Negroponte was also skillful at the high-risk game of setting corporate antagonists against each other, which he saw "a bit like sport fishing or a habit hard to kick."[68] As he explained:

> You can call a company's bluff. You can say: "You have two choices. You want to fund something for $300,000 a year, we'll give you an exclusive on that, and

we won't show it to anybody else. But when you come to the Lab, we're going to put blinders on you and take you up to the little room and show you the work we're doing for you, and then we will march you right out of the building again. There happens to be $6 million of other stuff going on in the building, but you're not going to see any of it." Or: "Give us the same $300,000; we'll give you complete access to the full $6 million package, and all you have to do is let us show other people what we're doing for you. Which would you prefer?" Everybody so far has said they prefer access to the full package.[69]

Although DARPA—the previous "closed world's" primary funding body—accounted for only 10 percent of the Media Lab's funding in 1987,[70] Negroponte managed to raise the requisite $45 million for the Wiesner Building by leveraging the voracious, corporate new media ecology against itself; providing an always "crazier" environment for the Lab's researchers and their corporate *voyeurs*.[71]

Part of this intrinsically specular, and spectacular, research modality was inherited from MIT's tradition of experimental, interdisciplinary work, as prototyped throughout the postwar period. But part of it was also in response to, and constitutive of, the very alliances the Lab had launched as a result of its programmatic ambitions of convergence, of which the funding strategy of paradoxical "aggregative disaggregation" was a central aspect, by necessity. As funding became decentralized and more competitive, so did the kinds of projects emerging from the Lab: each one more and more quirky, more and more "out there"—a veritable "zoo" (aka "vivarium") dutifully shepherded by Negroponte—in which cybernetic and digital turtles, as well as fish, termites, and public school children all played on display for the sake of corporations, acting as indirect insurance agents against the disruptive technological innovations—aka "capitalism"—they themselves were involved in producing. At the Media Lab, companies were able to anxiously (but legally) "peak" into each other's interests via their proxies—the researcher's simulated swarms—thus hedging the future's unexpected market anxieties for just another day, all the while heralding the dawn of the "Era of Decentralization."[72] Mobilizing the mortal fears embedded in such systems—technological, organic, and economic—the Media Lab found its own way of guiding the "invisible hand" of the market's unruly forces by taking potential funders on the ride of the capitalist sublime, and then offering them the hope of a further lease of artificial life, for a small donation:

How can you peer ten years along a technological trendline that might devour or starve your present cash cows? How can you explore the crossover technologies where entire new businesses are being born without becoming one of the stillborn? You read in the *Wall Street Journal* or the *Boston Globe* how former industrial backwater Massachusetts is booming, with unemployment down to 3.6 percent and a state budget surplus, and it's all being attributed to MIT. Then Negroponte shows up keynoting somewhere with video demos of MIT researchers test-piloting the information technologies at the edge of the possible, flying in formation around a pattern vague and shifting but emerging, hypnotic… and you buy in.[73]

Conclusion: Liberalism's "demo or die"[74]

Interestingly, this "buy in" necessitated, and was constituted by, an infrastructure of display: the Media Lab's corporate voyeurism, the *Wall Street Journal*, the *Boston Globe*, and the conference's stage upon which Negroponte would be "keynoting." On the one hand, this infrastructure of display was the instrument for a particular kind of invisibilization: the sudden disappearance of the jobs that had made Massachusetts into an "industrial backwater" in the first place—a disappearance congruous with the school defunding policies. This infrastructure of display sustained this elision as it became animated by Negroponte's "demos": the "emerging, hypnotic," and shifting flights in formation around vague patterns, Negroponte's "world-class salesmanship," together with the newspaper statistics and the stage lighting, were all designed to produce—in combination—a somewhat *magnificent* aesthetic effect in which, returning to Burke, one would be "so dazzled as to make it impossible to attend to that exact coherence and agreement of the allusions, which we should require on every other occasion."[75] Except for "magnificence," in this and every other occasion, became the rule.

In this infrastructural, mixed-media circuit of labs, newspapers, and demos, a profusion of images and numbers conjured the strong impression of the vital importance—literally a matter of life and death—of becoming a part of the system. The effects, credibility, and especially, the capital flowing through the circuit constituted a self-organizing, self-stabilizing organism—if you "bought in." Joining this quasi-organic infrastructure produced an economy of "sense"—an aesthetic experience on one level, but also in terms of the promise of absolute meaning it implied at another: higher returns; solving public education; inventing the future; saving the world.

Of course, literally joining or not joining—investing or not investing—was not actually important, the crucial aspect in this infrastructure of display was in the utter fact of being there: being able to see it happen—seeing your competitors (your "equals") seeing it happen—or imagining that something was indeed happening and which involved the imagined community of investors with which you identified. If, as Benedict Anderson argues, newspapers created the "imagined communities" of "print capitalism," TED demos and corporate boardroom presentations have constructed the imagined self-aggregations of our new media capitalism.[76]

Swarms, as endlessly repeated icons of this imagined community, are a performative operator in its techno-cultural infrastructure. Their performative power lies in their sheer capacity for circulation: they provide "sense" through their very circulation as images in the network's channels. In recent architecture, the swarm has become a pervasive cipher for a host of different attributes: adaptability, complexity, organicism, emergence, bottom-up nonhierarchical processes, nonregulation, self-organization, informality, or the idea of evolutionary design. Previous figures in architectural discourse that occupied this role of "sense"-operators, helping to construct their own historical-specific imaginaries, would include the "machine aesthetic" for early-twentieth-century functionalism, and the myth of the "primitive hut" for various kinds of classicist and ethno-structural rationalisms, to give just two examples.[77] The circulation of these particular imaginaries—through the circulation of printed books—constituted the infrastructural condition for the dissemination of their ideological and cultural charge.

With the Media Lab, the demo or keynote speech and the burgeoning aggregated digital-media-communications complex did the same for the imaginary of swarms—not just in terms of recirculation of these kinds of images and their affective qualities, but also in the technical-historical construction of the idea of human collectives as organic compounds of individual self-interests that would naturally self-organize and aggregate through capitalist collaboration and competition.

As we have seen, the Logo programming language prototyped a community of liberated, creative individuals (children), reproducing a subjectivity that programmatically elided the institutional, social, and political context in which it operated. Children they were in the 1970s, but this subjectivization model and apparatus clearly resurfaced in the shape of the Media Lab's researchers of the mid-1980s. With the same commitment to self-directed play-work and experiential curiosity, the students—many of whom were, not coincidentally, also actually interning for the Lab—were "unleashed" as the creative organic units of the greater Media Lab and the global corporate capitalist aggregate whole. Equipped with the necessary technical and cultural skills for competition in this new "information age" economy, the student-workers were clearly more prone to capitalist "self-organization" than actual political organization. In a more recent, shadow narrative to these developments, the Media Lab continues its proselytizing of capitalist individuation through the One Laptop per Child initiative—where, according to Negroponte's 2014 TED address, children the world over are given personal computers to use in their "primitive" environments.[78] Who needs schools when everyone can have a laptop?

Meanwhile, as this swarm-like model of production proliferates, on the other side of the river from the Media Lab remains Hennigan School, where Papert conducted his 1980s techno-pedagogical experiments, and which is still suffering the consequences of the defunding policies first begun by Reagan and then continued by other administrations.[79] This is a cruel irony for Papert's radical intentions, who had foreseen a total revolution in schooling through the rejection of the curriculum system, standardization and testing methods, and even teachers, in favor of a model where children could explore what they wanted and how they wanted, limited only by the technological environment available. In contrast, computing was incorporated into schools as another subject in itself, with its own curriculum, teacher, and means of testing. As Papert would note in the late 1980s, when he began his techno-pedagogical research of the 1970s, he saw the incorporation of the computer into the classroom as "an instrument of radical change. And then around about the middle of the 1980s this computer got into the hands of school administrations and the ministries and the commissioners of education, state education departments."[80] The result, Papert concluded, was that the computer had been "thoroughly assimilated to the way you do things in school"—institutional over-regulation had stifled the possibility of a much more radical change, where school might be entirely dissolved, if not utterly transformed. This transformation, Papert argued, was not historically or politically driven, but was the result of the new availability of tools—and whose full blossoming the education authorities were hampering. The situation, for Papert, was untenable:

> I'm saying that it is inconceivable that school as we've known it will continue. Inconceivable. And the reason why it's inconceivable is that little glimmer with my grandson who is used to finding knowledge when he wants to and can get it when he needs it, and can get in touch with other people and teachers, not because they are appointed by the state, but because he can contact them in some network somewhere. These children will not sit quietly in school and listen to a teacher give them predigested knowledge. I think that they will revolt.

Papert's interlocutor in this dialogue, the Brazilian educator and philosopher Paulo Freire, saw things quite differently, however. Describing Papert's argument as "metaphysical," not "scientific thinking," he argued that Papert was displacing the historical and political dimensions of the problems of learning to an ontological plane in which student-driven experiential learning (the ideal opposite of "being taught") would somehow happen automatically through the freedom of the computerized student from institutional structures that might get in the way. By contrast, Freire proposed:

> For me, the challenge is not to end school, but to change it completely and radically. (…) To me, the problem we face today is the correction of the mistakes of the [traditional school] that are not all didactic and not methodological mistakes but, indeed, ideological and political ones. Thus, what we must do is to change the world politically. It's the power that ought to be changed. In order to do this we shouldn't say that history is dead or that the [political] classes have disappeared. All this is just talking in order not to change the [school]. All these speeches of the new liberal perspective ideology are trying to preserve the [school as it is]. Nevertheless, in order for us to change the [school] we have to change the liberal speech.

Changing liberal discourse, Freire maintained, is more difficult and important than attempting institutional deregulation, via computers or otherwise, if one really wants to change education. The very idea that "freedom" should be defined as the private individual's "freedom from" institutions, fellow students, colleagues, teachers, or the State itself—rather than the positive formulation of "freedom for," that does not exclude the role of institutions and others—is at the root of liberal discourse.[81] Paradoxically, despite Papert's best intentions for producing a more liberating, rich and creative learning environment, his Logo project became a liberal techno-ideological apparatus that assumed, and thus reproduced, an idea of the virtuous subject as the one who acts according to his own self-interest, in total mastery of himself and his desire—sovereign.

The fact that these environments became the model for a whole economy of higher education—and thus, of the "information age" economy at large—is shown by the Media Lab's funding dynamics and its obsession with decoding the organic relations of collectives through artificial life projects, like simulated swarms. Each time these funding mechanisms and organicist spectacles were enacted—recursively, one for the other—the liberal creed of sovereignty and self-preservation was replayed again like a techno-cultural mantra. If the discourse of parametricism in architecture wants to generate something truly new, as Freire alerted, it is the very notion of sovereignty and organicist imaginaries of belonging (to selfhood, nature, culture, capital, decentralization, etc.) that it needs to sunder. And yet, it cannot do this from the same liberal perspective of attempting to "do away" with institutions, politics, or history. Recognizing that each "parametric" model always constructs a correlation between image and imaginary—fact to principle, autonomous student to functional society—*and taking a stance toward these relations and the forms of life and existence they recursively transform, annihilate or reproduce*, would begin to define a politics for parametricism.

Acknowledgment

The author would like to thank John Tyson for his valuable comments on an early version of this chapter. Thanks also to Felicity Scott, Reinhold Martin and Mabel Wilson for their Columbia seminars, from which many of these reflections emerged.

Notes

1. Reynolds's Master of Science thesis on procedural animation ("Computer Animation in the World of Actors and Scripts") was supervised by Nicholas Negroponte, the Architecture Machine Group's director. Reynolds also held a Bachelor of Science in Electrical Engineering and Computer Science from MIT (1975). See Reynolds's CV on his website http://www.red3d.com/cwr/resume.html [accessed November 12, 2014].
2. Sebastian Vehlken, "Zootechnologies: Swarming as a Cultural Technique," in *Theory, Culture, & Society, Special Issue: Cultural Techniques*, eds. Geoffrey Winthrop-Young, Ilinca Iurascu, & Jussi Parikka (London & New York: Sage Journals, November 2013), p. 30 (6).
3. For an example of an architectural swarm image, see this book's front cover. Image by ZHA.
4. Seymour Papert, "Teaching Children Thinking. Artificial Intelligence Memo Number 247" (MIT Cambridge, Artificial Intelligence Lab, 1971), p. 1. Available online at http://stager.org/articles/teachingchildren.html [accessed October 1, 2014].
5. Edmund Burke, *A Philosophical Enquiry into the Origin of Our Ideas of the Sublime and Beautiful*, ed. James T. Boulton (Notre Dame, IN: University of Notre Dame Press, 1958), part II, p. 78.
6. In sociological terms, this is the classic dialectical tension presented by Ferdinand Tonnies between the poles of "Gemeinschaft" (organic community) and "Gesellschaft" (artificial society).
7. Vehlken, "Zootechnologies: Swarming as a Cultural Technique," p. 122.
8. Craig W. Reynolds, "Flocks, Herds, and Schools: A Distributed Behavioral Mode," *Computer Graphics* Vol. 21, No. 4 (July 1987), p. 28.
9. Ibid., p. 30.
10. Ibid., p. 27.
11. Ibid., p. 30.

12. See, for instance, Kevin Mihata, "The Persistence of Emergence," in *Chaos, Complexity, and Sociology: Myths, Models, and Theories*, eds. Raymond A. Eve, Sara Horsfall, & Mary E. Lee (Thousand Oaks, CA: Sage Publications, 1997). Reynolds went on to work on 3D animated crowds for the special effects industry and even won an Academy Award in 1998. See Academy Awards 1998. Available online at http://www.imdb.com/event/ev0000003/1998 [accessed October 12, 2014].
13. Mihata, "The Persistence of Emergence," p. 31.
14. Ibid., p. 32.
15. Ibid., p. 35.
16. Alan M. Turing, "Computing Machinery and Intelligence," *Mind*, New Series, Vol. 59, No. 236 (October 1950), pp. 433–60.
17. See, for instance, Grey Walter's turtles, on which the Logo turtles were in fact based, via the work of Ivan Sutherland, another MIT graduate who worked on robotic turtles and computer graphics. See William Grey Walter, "An Imitation of Life," *Scientific American* Vol. 182, No. 5 (May 1950), pp. 42–45; Ivan E. Sutherland, *Sketchpad: A Man-Machine Graphical Communication System* (PhD dissertation) (Cambridge, MA: Department of Electrical Engineering, MIT, 1963); and Ivan E. Sutherland, "An Electro-Mechanical Model of Simple Animals," in *Computers and Automation*, ed. Ivan E. Sutherland (Cambridge, MA: MIT Press, February 1958); "Stability in Steering Control," *Electrical Engineering*, April 1960. Sutherland's personal correspondence with Walter can be found at http://cyberneticzoo.com/cyberneticanimals/1957-machinaversatilis-ivan-sutherland-american/ [accessed October 20, 2014].
18. Reynolds, "Flocks, Herds, and Schools," p. 27.
19. Seymour Papert, "Uses of Technology to Enhance Education. Artificial Intelligence Memo Number 298" (MIT Cambridge, Artificial Intelligence Lab, 1973), p. 100. Available online at http://hdl.handle.net/1721.1/6213 [accessed October 20, 2014].
20. In this sense, Papert's work was directly opposed to the "New Math" movement of the 1960s. Ibid., p. 46. See also Nathalie Sinclair, *The History of The Geometry Curriculum in the United States* (Charlotte, NC: Information Age Pub., 2008).
21. Papert, "Uses of Technology," p. 47.
22. Ibid., p. 19.
23. Seymour Papert, "Teaching Children to Be Mathematicians Vs. Teaching About Mathematics. Artificial Intelligence Memo Number 249" (MIT Cambridge, Artificial Intelligence Lab, 1971), p. 1. Available online at http://hdl.handle.net/1721.1/5837 [accessed October 20, 2014].
24. Seymour Papert, *Mindstorms: Children, Computers, and Powerful Ideas* (New York: Basic Books, 1980), p. 36.
25. Ibid., p. 19.
26. Seymour Papert, "Situating Constructionism," in *Constructionism*, eds. S. Papert & I. Harel (Norwood, NJ: Ablex Publishing Corporation, 1991), pp. 193–206. Available online at http://www.papert.org/articles/SituatingConstructionism.html [accessed October 1, 2014].

27. Papert, *Mindstorms*, p. 7.
28. Ibid., p. 206.
29. Ibid., p. 19.
30. Papert, "Teaching Children Thinking," p. 4.
31. Papert, *Mindstorms*, p. 63.
32. Ibid., p. 68.
33. Ibid., pp. 56–58.
34. Papert, "Teaching Children Thinking," p. 1.
35. See Cynthia Solomon's blog on the history of Logo at http://logothings.wikispaces.com/ [accessed October 5, 2014].
36. See the AI Lab's Logo Memos, available online at http://dspace.mit.edu/handle/1721.1/5460 [accessed October 5, 2014].
37. Papert, "Uses of Technology," p. 27.
38. Sinclair, *The History of the Geometry Curriculum*, p. 73.
39. Papert, *Mindstorms*, p. 64.
40. Papert, "Uses of Technology," p. 45.
41. Ibid., pp. 25–26.
42. Papert, "Teaching Children Thinking," p. 1.
43. See, Michel Foucault, *The Birth of Biopolitics: Lectures at the Collège De France, 1978–79* (Basingstoke: Palgrave Macmillan, 2008). In particular, see "Lecture 9: March 14, 1979," p. 226.
44. Papert, *Mindstorms*, pp. 71–74.
45. Ibid., p. 93.
46. Papert, "Uses of Technology," p. 68.
47. Ibid., p. 68.
48. Stewart Brand, *The Media Lab: Inventing the Future at MIT* (New York : Viking, 1987), p. 98.
49. See, for instance, Mathew Wisnioski, "Centrebeam: Art of the Environment," in *A Second Modernism: MIT, Architecture, and the "Techno-Social" Moment*, ed. Arindam Dutta (Cambridge, MA: The MIT Press, 2013), pp. 188–225.
50. See Wisnioski, "Centrebeam," 210–11; and Nicholas Negroponte, "The Origins of the Media Lab," in *Jerry Wiesner: Scientist, Statesman, Humanist: Memories and Memoirs*. Jerome B. Wiesner & Walter A. Rosenblith (Cambridge, MA: MIT Press, 2004), p. 150.
51. Negroponte, "Origins of the Media Lab," p. 151.
52. See, Nicholas Negroponte, *The Architecture Machine; Toward a More Human Environment* (Cambridge, MA: MIT Press, 1970); and, *Soft Architecture Machines* (Cambridge, MA: The MIT Press, 1975).
53. Brand, *The Media Lab*, 162. Quoting Minsky: "There was probably more freedom of research under ARPA than any other government agency, because they trusted the judgment of the people they supported. Why? Because they were us. For fifteen years the office down there was run by an ex-MIT person or equivalent. It was like having a patron."
54. Paul N. Edwards, *The Closed World: Computers and the Politics of Discourse in Cold War America* (Cambridge, MA: MIT Press, 1997).

55. Negroponte, "Origins of the Media Lab," p. 155.
56. See, Felicity Scott, "Discourse, Seek, Interact," in *A Second Modernism: MIT, Architecture, and the "Techno-Social" Moment*, ed. Arindam Dutta (Cambridge, MA: The MIT Press, 2013), pp. 392–93.
57. An important aesthetic inspiration for Negroponte had been Bernard Rudofsky's 1964 book, *Architecture Without Architects*, which portrayed quasi-organic aggregations of vernacular and "primitive" communal constructions around the world, as had been Moshe Safdie's 1967 aggregate construction of housing units ("Habitat" in Montreal). See, Brand, *The Media Lab*, 142; Nicholas Negroponte, "Toward a Theory of Architecture Machines," *Journal of Architectural Education* (1947–1974), Vol. 23, No. 2 (March 1969), pp. 9–12; and Bernard Rudofsky, *Architecture Without Architects, an Introduction to Nonpedigreed Architecture* (New York: Museum of Modern Art; distributed by Doubleday, Garden City, NY, 1964).
58. Negroponte quoted in, Scott, "Discourse, Seek, Interact," pp. 362, 365.
59. The National Commission on Excellence in Education, *A Nation at Risk: The Imperative for Educational Reform: A Report to the Nation and the Secretary of Education*, United States Department of Education (Washington, DC: National Commission on Excellence in Education, 1983), p. 5.
60. Ibid., p. 11.
61. Negroponte, "Origins of the Media Lab," p. 153.
62. Ithiel de Sola Pool, *Technologies of Freedom* (Cambridge, MA: Belknap Press, 1983).
63. Negroponte, "Origins of the Media Lab," p. 153.
64. Brand, *The Media Lab*, p. 8.
65. Ibid., p. 6.
66. Negroponte, "Origins of the Media Lab," p. 152.
67. Brand, *The Media Lab*, p. 156.
68. Negroponte, "Origins of the Media Lab," p. 155.
69. Negroponte quoted in: Brand, The Media Lab, p. 157.
70. Brand, *The Media Lab*, p. 163.
71. "[T]he comment I get the most these days from our industrial partners is 'be crazier.' In fact, the value to corporations is very much at the lunatic fringe, going out on a limb as much as possible, doing things they might not." Negroponte, "Origins of the Media Lab," p. 156.
72. Mitchel Resnick, *Turtles, Termites, and Traffic Jams: Explorations in Massively Parallel Microworlds* (Cambridge, MA: MIT Press, 1994), p. 7.
73. Brand, *The Media Lab*, p. 9.
74. Ibid., p. 3.
75. Burke, *A Philosophical Enquiry*, p. 78.
76. Benedict Anderson, *Imagined Communities: Reflections on the Origin and Spread of Nationalism* (London: Verso, 1991).
77. See, for instance, Reyner Banham, *Theory and Design in the First Machine Age*. (New York: Praeger, 1967); and Marc-Antoine Laugier, *An Essay on Architecture* (Los

Angeles: Hennessey & Ingalls, 1977 [1755]).
78. Nicholas Negroponte, "A 30-year history of the future," TED talk available online at https://www.ted.com/talks/nicholas_negroponte_a_30_year_history_of_the_future?language=en [accessed October 7, 2014].
79. As of 2014, Hennigan School was in the lowest 20 percent performing of schools of its category, according to official State statistics. See *Report on Teaching and Learning (RTL): School Report Card, School Year 2013–2014. James W. Hennigan School.* Available at http://www.bostonpublicschools.org/school/hennigan-elementary-school [accessed October 15, 2014].
80. Seymour Papert and Paolo Freire in conversation, "The Future of School," transcript available online at http://www.papert.org/articles/freire/freirePart1.html [accessed October 15, 2014]. All subsequent quotations are from this dialogue.
81. See Hannah Arendt, *"What is Freedom?,"* in *Between Past and Future: Eight Exercises in Political Thought* (New York: Viking Press, 1968).

Chapter 7

Breeding ideology: Parametricism and biological architecture

Christina Cogdell

The breeding ideology that envelops parametric architecture is but the most recent expression of a broader historical trend of biologizing and sexualizing architecture. For example, Contemporary Architecture Practice's design for a Residential Housing Tower in Dubai, with its seemingly intertwined, curvaceous, two-in-one tower formation, makes a racy intimate contribution to the city's skyline; the firm suggestively describes the "two contiguous buildings" as losing "their individual identity, fusing into a hyphenated formation."[1] The parametric software used for this design integrated such features as apartment floor-space and volume, ocean views, construction material properties, and associated production costs in order to maximize desirable features in light of specified parameters, including the budget.[2] Cerebral by contrast, Foreign Office Architects' (FOA) Evolution Chart (Figure 7.1) maps out the offspring of the firm's "DNA" classified in terms of families and species of buildings, each adapted to its individual environment. This evolutionary tree was created for the exhibition Breeding Architecture at London's Institute of Contemporary Arts in 2003 and published in the exhibition catalog *Phylogenesis: FOA's Ark*. In it, FOA describes their evolutionary computational design process: "This is not a simple bottom-up generation; it also requires a certain consistency that operates top-down from a practice's genetic potentials. Just as with horses and wines, there is a process in which successful traits are selected through experimentation and evolved by registering the results."[3]

Lest these examples seem to be radically innovative conceptions of architectural breeding, compare them with a few historical examples from the 1930s, specifically exhibitions at the 1939 New York World's Fair that combined male and female sexual symbolism as if this pairing were the basis of the

124　　　　　　　　　　　　　　　　　　　　　　　　The Politics of Parametricism

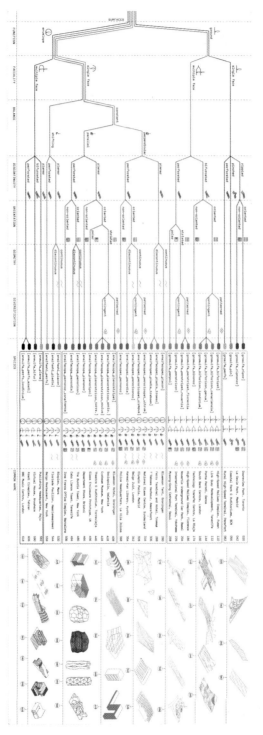

Figure 7.1 Foreign Office Architects, Classification System of FOA's offspring in *Phylogenesis: FOA's Ark* by Alejandro Zaera-Polo, Farshid Moussavi et al. (Barcelona: Actar, 2004), back page fold-out. Reproduced with permission from Alejandro Zaera-Polo, Farshid Moussavi.

fair's theme, "the world of tomorrow." I argue in my book *Eugenic Design* that these designs functioned within a context of widespread popular acceptance of eugenic ideology, which aimed to improve the genetic "fitness" of the national body through controlled breeding, enacted primarily through involuntary reproductive sterilization laws in over thirty states. "Fitness" was assumed to entail criteria of race, class, sexuality, and ability as demonstrated by the less-than-subtle innuendo of a Mendelian lesson about Color Inheritance in Guinea Pigs, displayed by the American Eugenics Society at state fairs in the 1920s. Using real taxidermied guinea pigs attached to a board, the display explained the proportions by which dominant or recessive traits appear in later generations as either "pure white," "pure black," "hybrid white," "hybrid black," etc. Its sister panel asserted: "Unfit human traits such as feeblemindedness, epilepsy, criminality, insanity, alcoholism, pauperism, and many others, run in families and are inherited in exactly the same way as color in guinea pigs." The panel then asked: "How long are we Americans to be so careful for the pedigree of our pigs and chicken and cattle, and then leave the *ancestry of our children* to chance, or to 'blind' sentiment?"[4] This goal of literally designing human evolution invoked an environmental design corollary: streamline and other modernist designers characterized their work as hastening the evolution of a pure, hygienic, future-perfect world.

Raymond Loewy's Evolution Charts from 1934 (Figure 7.2) depict everything from "a match to a city"—actually, from railcars to chairs to goblets—evolving toward the hygienic and eugenic streamlined ideal.[5] In contrast to the across-the-board *idealized* stylistic uniformity of the 1930s, FOA's evolutionary tree positions each of their firm's offspring as adapted to its particular site, with differentiated features selected accordingly. Owing in large part to the sociopolitical history of the six decades after World War II, including activist movements in civil, Lesbian, Gay, Bisexual, Transgender, and disability rights, twenty-first-century popular views of evolution like FOA's tend toward celebrating individuality and diversity. That being said, however, Patrik Schumacher's parametricist vision and language is much closer to Loewy's. His description, in his keynote address at The Politics of Parametricism conference, of the postmodern and deconstructed city as "garbage spill urbanism" echoes the twentieth-century interwar diatribes against immigration, a heterogeneous national social body, and the hybrid "primitive" ornamental aesthetics of Art Deco. His assumptions of inevitable "progress" resulting from the so-called march of cultural evolution are more Spencerian than Darwinian. Like Loewy, his and Zaha Hadid's visions of large-scale urban makeovers (such as their Kartal Pendik Masterplan for Istanbul) impose control on the city through "idealized" "optimized" elitist homogeneity. Parametricist

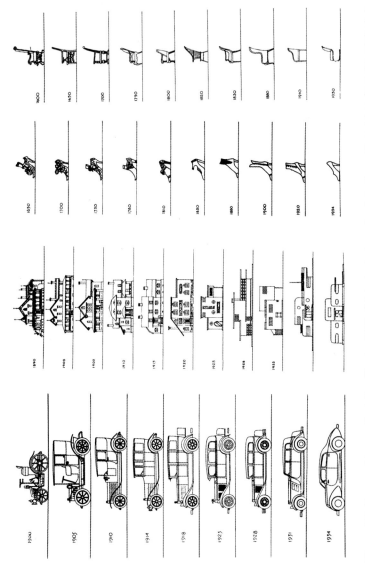

Figure 7.2 Raymond Loewy, Evolution Charts, 1934. Originals destroyed; prints of negatives taken from the originals in collection of the author. © Estate of Raymond Loewy.

urbanization such as theirs is perhaps aptly described as the twenty-first-century version of "streamlining."

Today's versions of architectural breeding thus differ from nineteenth- and twentieth-century evolutionary design in terms of their sociopolitical historical contexts, as well as in terms of revised scientific theories based upon new knowledge, often in combination with older lingering popular beliefs. While the DNA obsession popularized by the now-outdated central dogma of molecular biology lives on in advertising rhetoric and popular conceptions of genetic engineering, its oversimplified biological determinism is being tempered by new scientific understandings of epigenetics and biological systems complexity. These theoretical developments contextualize the gene as but one part of a very complex, environmentally sensitive system whose overall development and phenotypic outcomes are influenced by a myriad of environmental actors at a multitude of scales, from within the cell to the environment external to the organism. Furthermore, new technologies of scientific research and architectural production, made possible through computers, evolutionary computation, computer-controlled production, structural engineering, and laboratory techniques of tissue and genetic engineering, are enabling the simulation and construction of visually and structurally complex forms, be they *organic* or *inorganic* in their material basis.

These developments have led to the current materialist philosophical acceptance of the theoretical dissolution of the divide between "nature" and "culture." In the field of parametricist architecture, they have contributed to an exaggerated biologization of generative architecture. This may be due to the fact that many architects do not have training in biology and do not actively collaborate with scientists and genetic engineers, but yet they rely constantly on computers to generate parametric design, often using genetic algorithms (GAs) and evolutionary computation. The general tenor of their statements and writings, as well as those of their critics and curators, confuses the digital with the biological, the "cultural" with the "natural." For architects like FOA who are using biology and evolution as *analogies* to architecture and its design processes, this confusion is less troubling even if still difficult to parse. However, some architects today—such as Alberto Estévez and Dennis Dollens at the Universitat International de Catalunya in Barcelona (ESARQ) in Barcelona, and Marcos Cruz at the Bartlett School of Architecture in London—are proclaiming the imminence of a *biologically material* architecture, meaning one that is actually alive, grown from animal or plant cells propagated through tissue engineering on complex scaffolds. Even more dramatically, they propose using genetic engineering to alter DNA sequences in order to produce the phenotypic traits one desires through biological development. For example, various tissue-engineered works by the artists Oron Catts and Ionat Zurr (Figure 7.3) are upheld by curators and architects as the best examples we have so far of what a living architecture might become, despite the fact that a two-inch-tall "victimless leather" coat or "extra ear" is very different than a "semi-living" building.[6] The video *Econic Design* depicting Atlanta, Georgia, in the year 2108, made by Matthias Hollwich and Marc Kushner and their students, is at least closer in scale. It describes how genetically engineered kudzu vines become the "MEtreePOLIS," tearing down the grid of twentieth-century architecture and turning the urban landscape into an "off-the-grid" photosynthetically powered playground.

Figure 7.3 *Victimless Leather*, on display at the Museum of Modern Art, New York, in *Design and the Elastic Mind* (2007–2008). Photograph courtesy of the artists, Oron Catts and Ionat Zurr.

The actual realization of human-scale, biologically material architectures such as these—meaning, high-tech bioengineered architectures, and not just arborsculpture—would force architects and scientists alike to grapple with true biological developmental systems complexity, both within and outside of the computer and laboratory. Attempting this realization would bring to the fore the oversimplifications of today's rhetoric of architectural biologization. Even if one theoretically accepts the dissolution of nature and culture, in fact vast material distances separate digital DNA from deoxyribonucleic acid; evolutionary computation from actual biological evolution; architectural morphogenesis from the morphogenesis of developmental biological systems; simulation of the "environment" in scripted code from the actual complexity of the multiscalar environment of cells, organs, and organisms in dynamic interaction in the world outside the laboratory. In other words, despite the rhetorical similarities, crucial categorical, material, and disciplinary differences separate the knowledge and capabilities of computer programmers and computers, from those of scientists and technicians in laboratories, from the independent development of living organisms in the world beyond the laboratory, from architects and engineers making buildings. Even a cross-disciplinary collaboration of experts from these fields working together on a biologically material, living design project would not make a computer simulation of a biological system equivalent to its living counterpart. It is crucial that architects and curators acknowledge these differences and consider the sociopolitical implications of their breeding ideology, as manifested in both biologically *analogical* and *material* generative architecture.

The key design method underlying analogies between generative architectural design and biological evolution is that of evolutionary computation, one technique of which is the use of GAs (Figure 7.4), developed by John Holland in the 1970s. Soon afterward, John Frazer, working in association with the Architectural Association in London and the University of Ulster, applied GAs to problem-solving in architectural design, as described in his book *An Evolutionary Architecture*.[7] Originating between the 1950s and 1970s, evolutionary computation is theoretically based upon the central dogma of molecular biology as framed within the broader Neo-Darwinian context of the modern evolutionary synthesis. Centrally fixated upon the "gene" as the primary source of inheritance and trait determination, evolutionary computation and GAs enact a number of operations—including mutation, crossover, and selection based upon fitness criteria—upon a string of "genes" in order to rapidly generate a population of forms. Within the field of evolutionary biology, evolutionary computation aided the study of population genetics by simulating patterns of change over time

Column 1   Column 1b   Column 3b

Figure 7.4 John Frazer with Peter Graham, "Genetic Algorithms and the Evolution of Form," paper given at TISEA 3rd International Symposium on Electronic Art in Sydney, Australia, 1992. Image courtesy of John Frazer.

"In this experiment instead of James Gibb's elegant proportions controlling the elements of a classical Tuscan column, we substituted random numbers to generate populations of poorly proportioned columns (2 or 2b) which were controlled by a genetic algorithm to evolve (2 or 2a) either back to Gibb's original elegance (1 or 1b) or by intervening with user selection, to evolve an alternative to match the user's aesthetic preferences – in this case a very squat column to make the differences obvious (3 or 3b) Only about 10 generations were required to evolve radically different outcomes depending upon the selection criteria."

very rapidly compared to the slow rate of biological evolution. In architecture, it was and is a pragmatic tool that facilitates complex problem-solving and form-generation across multiple competing criteria. In both fields, evolutionary computation works through the process of optimizing fitness criteria defined by the programmer: offspring with optimal fitness ("elites") are selected and bred with each other to "improve" future generations, whereas offspring with less-than-optimal fitness are "killed" in order to remove them from the population's "gene pool."

This process is theoretically identical to that of eugenics, which used "positive" measures to encourage the so-called "fit" to breed often, and "negative" measures (such as involuntary sterilization, euthanasia, or genocide) to remove the genetic influence of the so-called "unfit" from various national "bodies." Eugenic sterilization laws remained on the books in the United States into the early 1980s, with over 70,000 victims, one-third of these in California. Even if many people, including scientists, consider eugenics to be a pseudoscience, it is clearly not a thing of the past; it has a continuous scientific and sociopolitical history up to the present. Recent examples of "negative" eugenics include the UK's Human Fertilization and Embryology Act (2008), which forbids the implantation of deaf embryos in women using IVF, even if they too are deaf. In California, between 2006 and 2010, female prisoners have been pressured to receive tubal ligations when receiving medical care in jail, especially during labor, which is coercive and illegal by federal law; almost 150 women have been sterilized without proper approvals.[8] "Positive" eugenics is now promoted through egg bank solicitation of eggs from tall athletic beautiful Ivy League women and talk of soon-to-be "designer babies." Eugenics—in its human, not digital, form—must therefore be included as part of the sociopolitical and cultural context of today's different expressions of breeding ideology.

When I refer to genetic algorithms as "eugenic algorithms" (EAs) in front of generative architects, I have encountered a number of reactions: dirty glances and an intentional redirection of conversation; discomfort with, but acknowledgment that GAs function as EAs; and active engagement with the key problematic assumption that the "designer" easily knows what optimization is, and debate over whether she has the right to impose it on "population design" in her disciplinary field (computer science, architecture, law, medicine, etc.). Clearly, optimizing a population of human beings differs from optimizing the fruit production of a plum tree, from using EAs to optimize the solar- and cost-efficiencies of a building or to simulate the optimal ecosystem balance between bunnies, foxes, and vegetation on a hypothetical island. Notice the categorical differences in the foregoing list in terms of materiality, enactment, enforcement, economy, and ethics. Our discomfort with EAs may stem not only from the legacy and ongoing discriminatory practice of eugenics, but also from the fact that in the different time periods when eugenic visions have become particularly popular, promoters grounded their claims upon scientific authority, theory, and practice, interpreted through the lens of current sociopolitical, economic, and cultural concerns. In other words, when eugenic breeding ideology is culturally prominent, the barriers ostensibly delineating the borders between authoritative scientific knowledge, popular interpretations of science, design theory and process, and sociopolitical policies were and are permeable and mutually influential. This alone gives us reason to pause and question the presumed accuracy and authority of the scientific foundations and sociopolitical dimensions of parametric design, in terms of its evolutionary and genetic conceptual foundations. For the broader technological, scientific, and sociopolitical contexts within which evolutionary architectures develop make a difference in the cultural capital that they carry and the sociopolitical agendas they support.

My initial concerns about the cultural import of widespread use of evolutionary computation and GAs have been both assuaged and troubled the more I have learned through my research. First, not all parametric design relies upon evolutionary computation or GAs, as is the case with simulated annealing and other types of complex problem-solving scripts. Second, as with violent video games, most people know the differences between actions completed in a computer and the same things done in real life; however, few people know the ongoing reality of eugenics, so they fail to consider real social correspondences to their digital processes. Third, the Neo-Darwinian biological determinism put forward by GAs is scientifically outdated. Architects thus relying upon GAs to lend credence to their rhetoric of architectural biologization need to update their approaches to evolutionary computation if they want to appear current: integrate epigenetics and systems feedback into the code, and seriously tackle how to include multiscalar "environmental" influences. Yet, just because the biological determinism promoted by GAs is scientifically passé does not mean that it therefore carries no cultural weight. Clearly, talk of "designer babies" and advertisements using DNA as symbols of superior product design both express and promote an overly simplistic biological determinism, giving reason for concern. Finally, as with eugenics from the 1930s, which appealed to Marxists, anarchists, and conservatives alike, merely espousing techniques of breeding "optimization" and "killing off" the unfit—be it within a computer, orchard, or nation—does not indicate or delimit one's politics, left, right, or center. We therefore have to ask of any particular project: What is being optimized, for whom, and which politics does it serve?

Evolutionary architecture originated in the 1970s out of John Frazer's left-leaning politics, aiming for tools that optimized designs to site-specific environmental factors and engaged users interactively in the configuration of structures suited to changing needs. Evolutionary architecture was a step in the direction of architectural artificial intelligence and artificial life. "Smart" buildings fed data from sensors would not only anticipate users' needs to save energy, but, in the absence of users, could alleviate their own "boredom" by reconfiguring themselves to better fit their current environmental conditions. Frazer acknowledged the appeal of a "grand unified theory" (GUT) of self-organizing complex systems (material, biological, social, economic, cultural), which would explain the emergence of order from chaos through the iterative bottom-up operation of simple rules at the local level of a system, informed by feedback. He optimistically hoped that evolutionary architecture would be used to ameliorate social and environmental systemic imbalances. However, he pessimistically acknowledged the possibility of its use to further disrupt and polarize social and environmental systems, owing to the human tendency to co-opt new scientific theories and technologies in order to dominate other people and the rest of nature.[9]

In the ensuing decades, theories of self-organizing complex systems and evolutionary computation have become the foundation for evolutionary economics and neoliberal globalization. Today's iconic structures of parametricist architecture are not publicly accessible affordable structures but instead are those that predominantly exist within privatized contexts of neoliberal economic development, owing to the high costs of engineering and constructing morphologically complex "starchitecture." This calls our attention to the rhetorical usefulness and democratic political symbolism of the phrase "bottom-up," which is applied as easily to anticapitalist protest movements like Occupy as to corporate competition in a deregulated market. Consider this in relation to Herzog & de Meuron's The Bird's Nest, designed and engineered by a global collaboration (almost certainly using evolutionary computation), built by investors developing Beijing for the 2008 Olympics under China's totalitarian political regime.[10] The visual imagery and bottom-up evolutionary framings—of both parametricist architecture and neoliberal economics—perform a significant amount of cultural work naturalizing (and "democratizing") different regimes of power. That this building was designed as the prime arena of international athletic competition, in which discourses of human genetic superiority, selection, and survival of the fittest always come into play, interweaves yet another strand of "breeding ideology" into this complex structure.

If parametricist architecture moves from the realm of organic analogy into organic materiality, in which buildings are grown from living plant or animal cells manipulated through technologies of tissue and genetic engineering, then the eugenic processes of selection and optimization at the heart of evolutionary computation become more tangible, metabolic, and problematic. Alberto Estévez, director of the Genetic Architectures Ph.D. program at the ESARQ, invokes "buildings whose walls and ceilings grow their own flesh and skin," covered with red or silver fur depending on a client's wishes. He is quick to point out that they will never need repainting. Yet somehow he, and Hollwich and Kushner in MEtreePOLIS, overlooks explaining how these buildings will be fed, where their waste will go, what happens when they change their shape or get sick or die—which will surely happen if they are grown outside of sterile glass bubbles—much less whether people want to reside or work inside living organisms. These critical questions are at the heart of Catts' and Zurr's artistic creations of "semi-living" coats and structures. The tissue-engineered "victimless leather" coat, on display at the Museum of Modern Art's exhibition Design and the Elastic Mind (2007), was made by seeding a coat-shaped scaffold with mouse cells; the cells divided and covered the scaffold, which then dissolved. The coat lived for only five weeks inside its sterile glass jar before the curator, Paola Antonelli, chose to "kill the coat" because it outgrew its original intended shape, becoming large and deformed with one arm falling off. Antonelli executed the coat by pulling the plug on the peristaltic pump; in other pieces by Catts and Zurr, the artists ask people to touch the semi-living structures, introducing bacteria that then kill the cells. Despite the serious points raised by Catts' and Zurr's work about what is entailed in actually growing, wearing, or residing in semi-living human-scale coats and buildings, Antonelli, Cruz, Estévez, and Dollens excitedly point to Catts' and Zurr's works as small-scale prototypes of the organic architectural future they foresee.

When I asked Estévez which genetic engineers he was working with in Spain, he said, "I need one. Do you know any?"[11] Perhaps his claim that "the architect of the future will no longer direct masons but genetic engineers" has made them reluctant to take him on as a collaborator.[12] For him, the biggest hurdle to growing living buildings is economics: "Who will be the new Christopher Columbus?" he asks. "And as with the discovery of America, as it is not something that must be invented—it is simply a question of time and money—the only thing remaining is to find the corresponding Queen Isabella to concede their [sic] personal jewels for such an enterprise." In my opinion, the scientific hurdles are much larger than Estévez realizes. His biologically deterministic design vision hinges upon the transformation of digitally designed DNA (expressed through 0s and 1s) into deoxyribonucleic acid through "robotic manipulation."

He assumes that a digital code symbolizing designed trait outcomes actually will present those outcomes phenotypically through organismal development, if a living cell receives a nuclear transfer with a somehow-corresponding manufactured ACTG sequence. This assumption fundamentally ignores the material, informatic, and systemic differences between digital evolutionary computation and actual biological processes, which require the materiality and interactive system complexity of cells, tissues, organisms, and their environments.[13] It also ties in to broader socioeconomic systems. Tissue engineering, for example, used to produce LabMeat or Catts' and Zurr's "victimless leather" jacket, relies upon 10 percent fetal calf serum (FCS) for the fluid that feeds the cells; FCS is extracted by the meat industry during the execution of pregnant cows. Tissue-engineered designs are decidedly not victimless or sustainable as they depend upon the cattle industry.[14]

To conclude, the breeding ideology that contextualizes today's evolutionary architecture and generative design is not just computational; it is also horticultural, agricultural, cultural, socioeconomic, and political. These various expressions of breeding ideology stem from a common design mentality, one that sees matter in its dual guises "nature" and "culture" as appropriate for and amenable to human manipulation and consumption. Near the end of *Econic Design*, Hollwich and Kushner aptly diagram the evolutionary development of this anthropocentric attitude; their black and white diagram first shows a tall tree with a small human next to it, then a small tree with a large human next to it, then a large human holding a small tree in his hand. I think their diagrammatic history, here, though, is tongue-in-cheek; after all, they chose genetically engineered kudzu to become the basis of the new metropolis. As a highly detrimental and extremely costly imported invasive species in its "natural" form, it aptly symbolizes the vast potential cost and damage, as well as the colonialist attitude, behind breeding ideology.[15] Estévez betrays this attitude so blatantly and arrogantly in his call for the new Christopher Columbus and Queen Isabella, Catts and Zurr more subtly when they ask visitors to kill their semi-living artworks through bacteria-laden touch.[16] The different materialities and ethics of these various breeding processes matter immensely. Digital zeros and ones are not the same thing as ACTG molecules, evolutionary computation not the same as biological evolution, cells not the same as people.[17] If architects choose to talk and act like these entities are equivalent through biological exaggerations, then the sociopolitical eugenic implications of evolutionary computation have to be taken very seriously. If optimizing populations by selecting elites to breed while killing off the unfit is routine professional practice, by ignoring the material and systemic differences between computers and living organisms, our ethics become quickly compromised. In other words, *architects do know the difference and should not act like they are the same thing*. At the same time, they should be cognizant that the relations between, transitions between, or transformations of these entities one into the other are systemically very complex, and they should understand these systems. Finally, eugenic beliefs and practices with regard to humans are ongoing in the present, and form part of the innuendo, implications, and even applications of evolutionary design strategy. Architecture and biology are inextricably sociopolitical.

Notes

1. From Contemporary Architecture Practice's website describing the Residential Housing Tower (also referred to as "Migrating Coastlines"). http://www.c-a-p.net/images_res_tower.html [accessed October 17, 2013].
2. Information from a lecture given by Ali Rahim at the University of Pennsylvania, fall 2008, in ARCH 741, when the author was a student in the course.
3. Foreign Office Architects, *Phylogenesis: FOA's Ark* (Barcelona: Actar, 2004), p. 11.
4. American Eugenics Society charts used at the Kansas Free Fair, in the collection of the American Philosophical Society, Philadelphia; discussed in Chapter 2, "Products or Bodies? Streamline Design and Eugenics as Applied Biology," in Christina Cogdell, *Eugenic Design: Streamlining America in the 1930s* (Philadelphia: University of Pennsylvania Press, 2004), pp. 33–83, and Figure 2.7.
5. Jeffrey Meikle, Chapter 6 "Everything from a Match to a City," in *Twentieth-Century Limited: Industrial Design in America, 1925–1939* (Philadelphia: Temple University Press, 2001).
6. Paola Antonelli, *Design and the Elastic Mind* (New York: Museum of Modern Art, 2007); Marcos Cruz & Neil Spiller, eds., *Neoplasmatic Design, AD* issue 78:6 (November/December 2008).
7. John Frazer, *An Evolutionary Architecture* (London: Architectural Association, 1995).
8. See Corey Johnson, "Female Inmates Sterilized in California Prisons Without Approval," Center for Investigative Reporting, July 7, 2013, available at http://cironline.org/reports/female-inmates-sterilized-california-prisons-without-approval-4917 [accessed October 26, 2013]. Corresponding stories ran in the *San Francisco Chronicle*, August 22, 2013, and *Sacramento Bee*, July 7, 2013. Also, North Carolina's legislature has just budgeted millions of dollars to pay living victims $50,000 each as compensation for the personal violation done to their bodies and lives by the state up until 1979. See Peter Hardin & Paul Lombardo, "North Carolina's Bold Model for Eugenics Compensation," *Richmond Times-Dispatch*, August 11, 2013, available at http://www.timesdispatch.com/opinion/their-opinion/north-carolina-s-bold-model-for-eugenics-compensation/article_10ed1912-b0ea-59b0-97d3-d69d6fff7203.html [accessed October 26, 2013].

9. Frazer, *An Evolutionary Architecture*, p. 21.
10. On the financing of the Beijing National Stadium, see http://www.chinadaily.com.cn/en/doc/2003-08/11/content_253698.htm [accessed October 26, 2013].
11. Conversation between the author and Alberto Estévez, after Estévez's presentation at the Association for Computer Aided Design in Architecture (ACADIA), October 2010, in New York City.
12. Alberto Estévez, *Genetic Architectures*, trans. Dulce Tienda (Barcelona and Santa Fe: SITES Books and Lumen, Inc., 2003), p. 17.
13. Estévez defines genetic architecture as "the fusion of cybernetic-digital resources with genetics, to continuously join the zeros and ones from the architectural drawing with those from the robotized manipulation of DNA, in order to organize the necessary genetic information that governs a habitable living being's natural growth, according to the designs previously prepared on the computer"; see his "Appendix for a Definition of Genetic Architecture and Other Related Terms," in *Genetic Architectures II* (Barcelona and Santa Fe, NM: SITES Books and Lumen, Inc., 2005), p. 78.
14. Antonelli, *Design and the Elastic Mind*, p. 115, oversaw the caption writing for the exhibition, which presented Catts' and Zurr's words at face value when they are meant to be sarcastic: "They argue that if the things we surround ourselves with every day can be both manufactured and living, growing entities, 'we will begin to take a more responsible attitude toward our environment and curb our destructive consumerism.'" On LabMeat, see Michael Hanlon, "Fake Meat: Is Science Fiction on the Verge of Becoming Fact?" *The Guardian*, June 22, 2012, available at http://www.theguardian.com/science/2012/jun/22/fake-meat-scientific-breakthroughs-research; Robert Ferris, "How to Make a Hamburger Without Killing the Cow," *Business Insider*, August 6, 2013, available at http://www.businessinsider.com/how-cultured-beef-is-made-2013-8; Maria Cheng, "No Cows Died to Make This Burger," *Jacksonville Florida Times-Union*, August 5, 2013, available at http://jacksonville.com/news/2013-08-05/story/no-cows-died-make-burger [accessed October 24, 2013].
15. I argue this more fully in "Tearing Down the Grid," *Design and Culture* Vol. 3, No. 1 (2011), pp. 75–84. Note that they also capitalized the ME and POLIS in MEtreePOLIS, making the text for tree smaller and centered between (i.e., the focus of) the human elements.
16. Estévez, "Genetic Architecture," in *Genetic Architectures*, p. 17; and, Alberto Estévez, "Biomorphic Architecture," in *Genetic Architectures II*, pp. 56–57.
17. When the author was conducting research at the Architectural Association (AA) in February 2011, I heard a conversation where a frustrated student in one of the evolutionary-computation-based graduate programs was asking a tutor when they were going to get to the biological part of the curriculum. She had come to the AA believing that she would be working in both the digital and biological arenas, and was considering dropping out of the program because she felt she had been misled as biological evolution was not part of the curriculum.

Chapter 8

Speculation, presumption, and assumption: The ideology of algebraic-to-parametric workspace
Matthew Poole

Where the technologies of parametric and algorithmic scripting are having an increasingly significant political effect is in the redefinition and reconfiguration of *work*; when and where work happens, and what work is or could be. Already such technologies are changing the structure of workflows for architects and other designers, and in turn they are changing the culture of design per se. But equally, these technologies are further facilitating and propagating an acceleration of post-Fordist labor models[1] and post-Taylorist management models.[2] These transformations are, in concert in a dynamic symbiotic relationship, significantly changing the spatial and temporal possibilities of what can be conceived of as "work."

Spatial flexibility is now found in many postindustrial jobs because of the extensive reach and complexity of digital communications technologies, and such work can be done remotely, "on the go" a long way from the static location of an office or factory. Simultaneously, this entails a temporal flexibility, where work can be undertaken at any time, leading to a fusion of leisure and work activity, where as long as one is connected to the internet one can work while doing anything else.

However, many theorists have argued that such flexibility actually increases the amount we work, albeit we have the illusion of working less. The Italian *Operaismo* theorists, including Paolo Virno, Maurizio Lazzarato, Christian Marazzi, and others, argue that work is sublimated into nonwork activities rather than being substantively diminished within these new work patterns.

These labor models have also seen dynamic automation introduced into work via cybernetic feedback mechanisms producing even greater spatial and temporal flexibility, but in order for these changes to be political *as such* there must be a significant reorientation and revision of what the technological means of production can do to the relations of production[3] *and* the relations of distribution[4] of that which work produces *and* through which work is produced. We have to rethink what "productivity" means in an age where parametric and algorithmic scripting technologies are the fastest, most efficient, most extensive, most "productive," *and* most dominant work tools in a global economy whose most precious commodity is "information."

In Jean-François Lyotard's influential book *The Postmodern Condition: A Report on Knowledge*, he demonstrates that knowledge is fundamentally changing in postindustrialized societies from a condition of "know-how" (*savoir-faire*—and thus, a knowing of how to work; i.e., how to qualitatively transform things in the world) to a condition of "what is" (*savoir ce qui est*—a constructible matrix of information about what the world is; i.e., a quantitative calculation of the world's *value*). "Knowledge" shifts from a condition of skill acquisition, with which to transform the world to a future state, to a condition of hoarding and exchanging of information, with which to evaluate the world and to measure "what is" in any presumed present. This has a fundamental effect upon valuing knowledge because knowledge as information takes on the operational form of a commodity and the relationship of the "knower" to knowledge becomes increasingly exteriorized. Lyotard explains:

> The nature of knowledge cannot survive unchanged within this context of general [digital electronic computational] transformation. It can fit into the new channels, and become operational, only if learning is translated into quantities of information. We can predict that anything in the constituted body of knowledge that is not translatable in this way will be abandoned and that the direction of new research will be dictated by the possibility of its eventual results being translatable into computer language. (…) Along with the hegemony of computers comes a certain logic, and therefore a certain set of prescriptions determining which statements are accepted as "knowledge" statements.[5]

Lyotard explains why this is so important, and is worth quoting at length as it is crucial for understanding the social and political import of post-Fordism:

> Knowledge is and will be produced in order to be sold, it is and will be consumed in order to be valorised in a new production: in both cases, the goal is exchange. Knowledge ceases to be an end in itself, it loses its "use value". (…) Knowledge in the form of an informational commodity indispensible to productive power is already, and will continue to be, a major – perhaps *the*

major – stake in the worldwide competition for power. It is conceivable that the nation-states will one day fight for control of information, just as they battled in the past for control over territory, (...) raw materials and cheap labour. (...) [T]he mercantilization of knowledge is bound to affect the privilege the nation-states have enjoyed, and still enjoy, with respect to the production and distribution of learning. The notion that learning falls within the purview of the State, as the brain or mind of society, will become more and more outdated with the increasing strength of the opposite principle, according to which society exists and progresses only if the messages circulating within it are rich in information and easy to decode. The ideology of communicational "transparency," which goes hand in hand with the commercialization of knowledge, will begin to perceive the State as a factor of opacity and "noise." (...) The question threatens to become even more thorny with the development of computer technology and telematics.[6]

This shift from knowledge as *savoir-faire* to "information as commodity" has a fundamental effect on work in turn because *the acquisition of knowledge in order to work* shifts to a position where *the acquisition of knowledge as information* becomes the primary concern and operation *of* work. Work becomes the very operation of the accumulation, funneling, and ordering of information.

Parametrics, however, has the possibility of treating both the qualitative condition of knowledge (as *savoir-faire*) and the quantitative condition of information as commodity (*savoir ce qui est*) simultaneously, and potentially presents tools with which to overcome the principal problems that Lyotard outlines.

The automation of design processes, data management, and information modeling via digital electronic tools is no doubt transforming the principal role of the designer. Design work is rapidly moving from formats where calculations, analyses, and schematizations are undertaken by human minds toward greater usage of digital electronic computation that makes such decisions automatically. Where previously the designer could have been conceived of as a skilled creative problem-solver with a specific evolving style, a signature aesthetic palette and a carefully constructed genealogy from the history of design, each embodying and conjugating various epistemological, aesthetic, formal and even moral and ethical factors, now the designer can delegate such definitional characteristics of their work to the software. Software increasingly makes more choices and decisions on such matters, and indeed the very idea of style in design is becoming anachronistic, because the final shape, form, and function of designed materials is determined by and contingent upon the dynamic parameters of the information flows that both inform and form the design process. Design need no longer produce schematics of production processes that employ fixed determinate factors or determined parameters for the production mechanics of finished "things" as such, but rather it can provide complex conduits that have the capacity to merge immense, fast-moving, constantly changing and diverse information streams through which rates of changing conditions (dynamic parameters) for designed materials and production processes can be tracked, mapped, and manipulated, leaving the designer free to operate more like a researcher or consultant, or even a kind of guru/quasi-philosopher figure.

Such innovations leading to the transformation of design have many positive aspects; however, there are also significant consequences for design per se, for relations of production and distribution, *and* for the relationship of work to information and knowledge in general. Such changes in the role of design introduce a character of pragmatic naturalism into the design ethos, leading to such now well-known and widely accepted claims as those of Patrik Schumacher that

> Architecture finds itself at the mid-point of an ongoing cycle of innovative adaptation – retooling the discipline and adapting the architectural and urban environment to the socio-economic era of post-Fordism. (…) organising and articulating the increased complexity of post-Fordist society. The task is to develop an architectural and urban repertoire that is geared up to create complex, polycentric urban and architectural fields, which are densely layered and continuously differentiated. Contemporary avant-garde architecture is addressing the demand for an increased level of articulated complexity by means of retooling its methods on the basis of parametric design systems.[7]

Schumacher's task for architecture and urbanism is to "redesign" or evolve these design disciplines in the image of the already given socioeconomic system of post-Fordism: to react to the increased levels of complexity, interactivity, and "inter-iteration" that post-Fordism affords to socioeconomic systems (i.e., the world of work) to "address the demand," as he puts it. However, he also goes on to say:

> Contemporary architecture aims to construct new logics – the logic of fields – that gear-up to organize and articulate the new level of dynamism and complexity of contemporary society.[8]

He proposes here that architecture and urbanism can and should purposively construct "new logics," or conditioning principles, that are to be projected onto or into the new levels of dynamism and complexity of contemporary society's methods of working in order to organize and articulate such levels better, and to make "work" evermore organized and synchronized, and thus more "productively efficient."

However, an important contradiction arises here: according to Schumacher, the new principal task of architecture and urbanism is to be both *reactive* and *proactive*—they must react to address the demand described, yet also proactively construct new logics to assist in the propagation of such a demand. Importantly, this contradiction is not at all alien to the logics of post-Fordism, contemporary capitalist economies and neoliberal politico-economics because it powerfully describes and delineates the key tenets of how work is recharacterized within post-Fordist labor modeling, post-Taylorist management structures and neoliberalism; that work can be *any* activity, including nonwork activities, and still be work per se.

The complexity of Schumacher's demand for the design industries, and the complexity of the wider ramifications of the increasingly naturalized contradiction outlined, is worthy of note because it presents us with a clear example of how ideology is an essential yet obscure part of the neo-pragmatism embodied by the parametricist paradigm, the new paradigms of post-Fordist labor models and the post-Taylorist management paradigms and their burgeoning respective and overlapping *ethae*, and also how it is implicitly aligned with the tenets and goals of neoliberalism.

Furthermore, Schumacher also states that this new pragmatics of parametricist design is analogous to scientific research paradigms, in that "styles are design research programs."[9] However, the caveat "analogous to" introduces some ambiguity into what is precisely meant here. To overcome this gap in the rationality of the thesis, Schumacher proposes *heuristics* as a pragmatic methodological necessity:

> The programme/style consists of methodological rules: some tell us what paths of research to avoid (negative heuristics), and others what paths to pursue (positive heuristics). The *negative heuristics* formulates strictures that prevent the relapse into old patterns that are not fully consistent with the core, and the *positive heuristics* offers guiding principles and preferred techniques that allow the work to fast-forward in one direction. The defining heuristics of parametricism are fully reflected in the taboos and dogmas of contemporary avant-garde design culture:
>
> *Negative heuristics*: avoid familiar typologies, avoid platonic/hermetic objects, avoid clear-cut zones/territories, avoid repetition, avoid straight lines, avoid right angles, avoid corners, (…) and most importantly: *do not add or subtract without elaborate interarticulations*.

Positive heuristics: interarticulate, hyberdize, morph, deterritorialize, deform, iterate, use splines, nurbs, generative components, script rather than model, (…).[10]

Here, we see how divisive the analogy in Schumacher's statement is, as the categorization of heuristic types into positive and negative methods tacitly introduces ideology to the program he outlines. A genuine nonidealist materialist scientific methodology would never presume to privilege one mode of heuristics over another. Instead, we see Schumacher's version of Parametricism in these sentences forcefully proscribing heuristic methods that produce results that are either *liked* more than others and/or are producing results that are more *likely* to smoothly dovetail with the demands of the market (i.e., with the demands of the hidden and attenuated ideology of late-stage postindustrial capitalism and neoliberal politics). Also, ironically, Schumacher's categorization of heuristics is itself heuristic, in that it is essentially a "rule of thumb."

Parametricism as described by Schumacher (and parametrics in general) is, however, not responsible for the origination of post-Fordism and post-Taylorism. These have evolved from the inherent contradictory logics of late-stage capitalism per se. The contradictory logics of the matrix of ideologies that we live within in postindustrialized societies occur because of our adherence to the principles of both neoliberal capitalism and liberal humanist democracy in tandem, largely via the liberalization and deregulation of markets' behaviors alongside tighter regulation of citizens'/subjects' behaviors via the rule of law. However, parametricism and parametrics *are* having an accelerative effect upon the development and extension of these new modes of working, and the contradiction that is presented by Schumacher above, plus the contradiction that parametricism and the current deployment of parametrics in general (which makes almost all industries evermore efficiently productive within the given accepted socioeconomic matrix of late-stage postindustrial capitalism) may also help us think further as to how parametric and algorithmic scripting technologies might be deployed to resolve the contradiction itself to overcome the many and varied stultifications that these developments in late-stage postindustrial capitalism necessarily produce.

To move further with this argument, we have to consider two key areas. First, we will explore what work has been and the changing spatiotemporal conditions of work within post-Fordism. Secondly, we will look at the ramifications of the paradigm of parametrics and how the core mathematics of this generative system of spatiotemporal morphology operates versus more basic arithmetic and algebraic conceptions of space and time, to see how parametrics accelerates the changing spatiotemporal conditions of work at the level of a changing conception of space and time.

There are many historical examples that describe and analyze work as a negative category of activity, determining it to have a detrimental effect on human well-being. For example, in this dialogue between Socrates and Critobulus, we see Socrates stating that labor is powerfully injurious to both the physical and mental health of a person, as well as to the overall health of the civic corpus:

> (...) the true gentleman should practise no mechanical arts; but rather agriculture and war (...) for not only are the arts which we call mechanical generally held in bad repute, (...) they are injurious to the bodily health of workmen and overseers, in that they compel them to be seated and indoors, (...) all the day before a fire. And when the body grows effeminate, the mind also becomes weaker and weaker. And the mechanical arts (...) will not let men unite with them care for friends and State, so that men engaged in them must ever appear to be both bad friends and poor defenders of their country.[11]

That a life spent working at a specific task that is strenuous, separated from others, is observed to have powerful deleterious effects on the body and mind of the worker, which then in aggregate "infects" the metaphorical body and mind of the State, is an observation almost entirely ignored by eighteenth-, nineteenth-, and early-twentieth-century industrialists. Putting into practice ideas such as Adam Smith's celebration of the division of labor and Charles Babbage's promotion of machinic production, exacerbating the refinement of divisions of labor by relegating the worker to "machine operator" or component part of the machine, modern industrial production grew at a massive rate through the last two and a half centuries as processes of work were evermore specialized, albeit that such processes of work were directly proportionately evermore de-skilled as specific tasks became more particular and repetitive.

The first sentence of Adam Smith's 1776 magnum opus, *An Inquiry into the Nature and Causes of the Wealth of Nations*, reads:

> The greatest improvement in the productive powers of labour, and the greater part of the skill, dexterity, and judgment with which it is anywhere directed, or applied, seem to have been the effects of the division of labour.[12]

Similarly, Babbage's 1832 masterwork, *On the Economy of Machinery and Manufactures*, celebrates the achievements of divisions of labor as quasi-natural and entirely pragmatic:

> Perhaps the most important principle on which the economy of a manufacture depends is the division of labour (...). The first application of this principle must have been made in a very early stage of society, for it must soon have

been apparent, that a larger number of comforts and conveniences could be acquired by each individual, if one man restricted his occupation to the art of making bows, another to that of building houses, a third boats, (…). This division of labour into trades was not, however, the result of an opinion that the general riches of the community would be increased by such an arrangement; but it must have arisen from the circumstance of each individual so employed discovering that he himself could thus make a greater profit of his labour than by pursuing more varied occupations. Society must have made considerable advances before this principle could have been carried into the workshop; for it is only in countries which have attained a high degree of civilization, and in articles in which there is a great competition amongst the producers, that the most perfect system of the division of labour is to be observed.[13]

However, the de-skilling of workers is famously criticized by Karl Marx in his analysis of these models of working,[14] which we now call Fordist labor models, as an egregious alienating process, whereby the sheer particularity and repetition of the work alienates workers almost entirely from the overall production process. Marx believed that the refinement of divisions of labor in factories led to the workers becoming "depressed spiritually and physically to the condition of a machine." That the worker is literally and metaphorically separated from the products of their work—and hence their relationship to the products of their work is abstracted, becoming evermore abstract, albeit that their work itself becomes every more grindingly material—also means that their relationship to other products that they encounter outside of work becomes equally abstract. This occurs primarily because the value of such products encountered cannot be calculated on any grounded scale, so it is only possible to accept the literal "price" of the goods rather than any wider more complex "value" of the goods in question, albeit that this price is an abstraction so complicated and abstracted in its relation to the labor process so as to be both virtually and actually an ineffable social index.[15]

The result of this is that social relations per se are rendered as ineffable—i.e., ideological—leading to a creeping alienation of all members of a society from one another and from processes of production, including the production of society itself. Marx was well aware of this early on in his analysis of industrial capitalism, and he shared the view that the division of labor was bound up in a symbiotic relationship with the literal and metaphoric division of space and time:

> The foundation of every division of labor that is well developed, and brought about by the exchange of commodities, is the separation between town and country. It may be said that the whole economic history of society is summed up in the movement of this antithesis.[16]

Marx and his close collaborator Friedrich Engels, however, did not believe that labor per se was a negative category of activity. The critique of labor found in their canonical writings is specifically focused on the conditions and relations of labor under the coextensive systems of capitalist production and liberal humanist ideology. In the unfinished essay *The Part Played by Labour in the Transition from Ape to Man*, Engels writes:

> Labour is the source of all wealth, the political economists assert. And it really is the source – next to nature, which supplies it with the material that it converts into wealth. But it is even infinitely more than this. It is the prime basic condition for all human existence, and this to such an extent that, in a sense, we have to say that labour created man himself.[17]

He goes on to describe a largely Darwinian schema of human evolutionary development that details how the necessity of working together to survive and thrive led to the physical development of evermore dexterous hands, brains, and even the larynx of the *Homo sapiens*. He describes how designing and using tools allowed early humans to work together, to develop speech and then complex conceptual representational schematizations of the world (and hence of time and space), ultimately leading to the complex organ that is *society*. Equally, as Engels describes a former naturalism where labor is an intrinsically positive force for human good, we see Marx developing a positive conception of labor outside of a bourgeois capitalist culture that will universally benefit all humankind:

> In a higher phase of communist society, after the enslaving subordination of individuals under the division of labor, and therewith also the antithesis between mental and physical labor, has vanished, after labor has become not merely a means to live but has become itself the primary necessity of life, after the productive forces have also increased with the all-round development of the individual, and all the springs of cooperative wealth flow more abundantly - only then can the narrow horizon of bourgeois right be fully left behind and society inscribe on its banners: from each according to his ability, to each according to his needs.[18]

Here, Marx's famous dictum unambiguously describes a division of labor, but one that is no longer defined and restricted by the conditions of a capitalist economy and its necessary contradictory ineffables produced by the reification of ideological ideals, but instead is determined and calculated *scientifically*, as a nonidealist materialism.

In considering different models of labor, we must consider their effect on conceptions and perceptions of time and space. Times and spaces of work under a Fordist division of labor are tightly regulated, controlled, and conditioned as the industrial "machine" seeks ever-greater efficiency within a finite fixed set of spatiotemporal parameters. The principal factor that necessitates this tightening is that the labor force, like any part of a machine, needs to be maintained and refreshed; i.e., reproduced—the workers must be able to work when *at work*, and so the potentiality of the worker's labor must be carefully managed both at work, at home, and at other times and spaces of living. A worker in the Fordist labor paradigm cannot work twenty-four hours a day, seven days a week because they would be physically exhausted, having no time to eat, to rest, to sleep, nor to renew their psychological vigor for working (e.g., by engaging in nonwork activities; i.e., leisure). The Fordist paradigm of work therefore, quite obviously, has a finite capacity for productivity because ultimately there is a finite limit to the tightening of efficiency of the production process's relation to the labor force. The potentiality for laboring that any given worker has in a system that places spatial and temporal boundaries upon work has a narrow and relatively easily calculated limit condition.

The spatiotemporal limit condition here is specifically defined by an arithmetic relationship between the "moving parts" of the Fordist paradigm, where there is a direct correlation of spatial and temporal manipulation to the value of the work undertaken. In other words, if the resources of space and time can be minimized between work operations, within a given time frame, then maximum efficiency can be achieved within the given fixed parameters of, say, human endurance, labor force size and strength, the length of a day, etc. Calculations of efficiency in this production paradigm only involve simple arithmetic operations, including addition, subtraction, multiplication, and division. You can add, subtract, multiply, or divide the quantity of factors that provide productive force, within the fixed parameters of maximum efficiency, to achieve desirable effects in the production process: more or fewer workers could be employed, they could work longer or shorter days, they could work faster or slower, etc. Equally, the space taken up by operations could be arithmetically increased or decreased accordingly. In a purely Fordist model of labor, the production processes are related only by variables that are governed in a strictly arithmetic spatiotemporal matrix of possibilities.

However, this description of a "purely" Fordist production paradigm is only hypothetical because there are always extrinsic variables occurring outside of the material production process conditioning its intrinsic factors. The market into which the products of this production paradigm must necessarily be inserted exerts some of these extrinsic forces. Similarly, social factors such as traditions, customs, religions, laws, or government policies have direct or indirect influence effecting force upon the conditions and parameters of production. Such forces, whether market factors or social factors, apply regulatory or constraining effects upon the processes and relations of production. As such, an actually existing Fordist labor model working in a market-based economy within an actual society—with group and individual identities, communities, histories, ideologies, judiciaries and policies, etc.—is subject to a system of dynamically related variables, and therefore its relations of production conform to a more complex algebraic relationality, where spatiotemporal relations are not only immediately interdependent but are also dependent upon the more ineffable qualities of the ideas, habits, wants, desires, and mores of the social relations of the society that (re)produce the complexities of the market itself.

To give a simple example, in a market-based economy, where production is both a material and an immaterial activity, even though there is a spatiotemporal limit condition to the actual material production process of goods and services, this *does not* mean that there is necessarily such a firm limiting of "value" (or we might say "productivity"; as *profit* becomes the measure of productivity, unlike in the hypothetical purely Fordist model where actually producing the products is the productivity) for goods and services, and hence there is not such a firm limiting of profits that can be made. In any market-based economy, the production of value of goods and services (i.e., the valorization of goods and services as commodities) is only partially related to the material production process, and hence only partially related to labor forces and their value. As is obvious, the powerful forces of supply and demand, which can be and are tightly regulated by the producers and distributors of goods and services, can inflate or deflate the perception of "value" by manipulating the demand, via marketing and advertising for example, and by restricting the supply of the given goods or services. Again, we see the introduction of ineffables into the relationship of work and value, or in other words work and "production efficiency." Where there is an abyssal void between actual material production and the perception of value, which contributes massively to the production of value per se, we see ideology filling the gap with ineffables such as quality, authenticity, innovation, novelty, fashion, etc., all of which produce *demand*, which can increase value, and hence productivity (i.e., *profits*).

Such ineffable ideological vectors no doubt have a material effect upon the production process as they have a direct effect upon the value of any goods or services, changing the status of supply and demand in that market, but they also enhance and accelerate the alienation of members of a given society under such condition. This is because not only are workers alienated from their labor both intensively and extensively, but as consumers of goods and services, the values of which are determined and driven by such speculative ineffables, they are then doubly alienated at the level of consumption. This partial loosening of the spatiotemporal limit condition in the production of value via ideological factors is defined and produced in an algebraic relationship between the production of value per se and the relations of production of goods and services, *and* relations of distribution of such goods and services. In fact, the alienation described above is a necessary condition for this algebraic spatiotemporal relation with work to actually *work*.

This algebraic relation works by introducing telic variables into the production paradigm. The *telae* (literally, far away goals; and more often than not practicably unattainable goals, such as "contentment" or even "happiness") are embodied in what the relations of distribution "promise" will be delivered by the products. The product is "loaded" as a sign promising fulfillment of a goal, which *in its most full form* is ultimately unachievable (hence, it is a *telos*), but nonetheless exerts a powerful force upon you to aim for such achievement. The *telae*, or promises, are directly *and* indirectly the factors that alienate us from the product itself, as the burden of signifier of the *telos* is too great, and hence the product itself is necessarily doomed to fail in the fulfillment of the promise made in its name.

The telic variables in the algebraic relation of production in actually existing Fordist production paradigms are therefore analogous to Platonic forms; pure essential "types." These pure (metaphorically) geometric forms drive the function of the algebraic relations in this paradigm. These forms are a type of variable that are fixed, in as much as they are a "type" with a specific function or shape, and as such they "drive" the dimensions of the algebraic relation therein. Such "types" have to be invented, dreamt up, reinvented, and innovated constantly within this production paradigm; hence the almost hysterical obsession with "the new" and newness that almost all industries have. This is quite different to the post-Fordist paradigm, which also has algebraic relations of production but whose algebraic relations of production are parametric, meaning that the dimensions of the relations drive the geometry or shape of the function, making redundant the necessity of "types," newness, invention, and innovation as such.

The apotheosis and limit condition of efficiency of actually existing Fordist production processes can be seen in what is known as Toyotist "lean production."[19] Here, goods are effectively sold before they have been produced, stemming from the innovation of "just-in-time manufacturing" that works extremely quickly using a matrix of preexisting configurations of predesigned and premade parts to produce semi-bespoke products most efficiently to fulfill an already measured demand before it ebbs away. This final innovation in the

Fordist production paradigm tips work and production into a new realm, which we now call post-Fordism.

The principal change in orientation of the production paradigm and its relations of production is that material production is relegated to a secondary activity, whereas the production of the *demand* for the material goods or services (i.e., the production of the market, and the necessary subjects and subjectivities therein) becomes the principal activity of industry.[20] This relatively recent innovation in the relations of production, the relations of distribution, social relations, and the relations of capital transforms the focus of production from the goods and services to "the production of the consumer" (specifically, the production of the subjectivity of the consumer) because the consumer and their needs, wants, and desires become the practicably calculable and (re)producible target of industrial production. If, as has happened, the actual manufacturing process of goods and services reaches a condition that is so quick, efficient, multiply-articulate, nimble and thus reactive, then the primary role of operations of an industrial producer becomes one whereby all that is left to "control" and "activate" is the multiple and diverse subjectivities of its presumed market. The primary work of industry then becomes preemptively presumptive, actively presuming to understand the complex subjective nature of the market (and its constituent members) and assuming the role of affective modulator of such subjectivities therein. In other words, Toyotism as a production paradigm assumes the position of a subject from its presumed market base (i.e., the consumer's subjectivity) prior to the actual subject of the presumed market base acting (i.e., before the consumer makes a decision to buy any particular goods or services), or even existing (i.e., before the consumer even knows that they are the kind of person who might take that decision). The character of the work in this production model then begins to take on the dual and contradictory character that we see in full-blown post-Fordism; that it is both simultaneously reactive *and* active.

As we observe the rapidly developing conditions of work under post-Fordist labor models, we can see how the limit conditions of previous models of work under capitalist economies are being transformed and expanded toward what appears to be an infinite field of possibilities for production, productivity, and times and spaces of work. This appearance of an infinite field of possibilities for work (i.e., the production of subjectivities) under post-Fordism occurs because its products are, and its principal labor modality is, immaterial.

"Immaterial labor" is labor that can occur anywhere and at any time because it is the work of modulating affects that determines the relation of a given subject to the hegemonic forces (in our case, "capital") within the society in which it operates, and thus is the work of affecting subjectivity and simultaneously reconstructing the possibility of the subject's relation to the central principle that drives the hegemonic forces (i.e., the accumulation of capital centripetally toward a given subject). Maurizio Lazzarato has defined two key characteristics of immaterial labor as:

> On the one hand, [regarding] the "informational content" of the commodity, it refers directly to the changes taking place in workers' labour processes (...) where the skills involved in direct labour are increasingly skills involving cybernetics and computer control (and horizontal and vertical communication). On the other hand, as regards the activity that produces the "cultural content" of the commodity, immaterial labour involves a series of activities that are not normally recognized as "work" – in other words, the kinds of activities involved in defining and fixing cultural and artistic standards, fashions, tastes, consumer norms, and, more strategically, public opinion.[21]

What is important in the relation of the subject toward the capital relation here is that an exponentially spiraling asymptotic ("fractal") relation occurs as the subject of immaterial labor "labors immaterially" to produce its subjectivity, which is of course the expression of the capital relation to which it is constantly (re-)building a relation. If we take this to its logical conclusion, the subject becomes nothing more than the qualities of its immaterial labor, which is the matrix of characteristics of the (dynamic) capital relation. Thus, the subject's subjectivity becomes nothing more than the expression of the capital relation's dynamic involution, of which it is but a dynamic malleable part. The subject, in this instance, becomes what we could call simultaneously a hyper-subject *and* a hypo-subject (i.e., it is simultaneously both more *and* less than a subject, because within the capital relation it appears to have a greater amount of agency than, say, in its relation to a liberal humanist democratic system of power, but without the capital relation it has a severely diminished agency in its relation to a liberal humanist democratic system of power). This is the principal contradictory condition of the hidden and attenuated ideology of late-stage postindustrial capitalism.

Immaterial labor not only pervades any space and time of living, but it pervades consciousness at the level of subjectivity in ways that are not definitively calculable spatiotemporally (i.e., they are speculative and "fractally" involuted). If we accept that immaterial production is now the fastest growing sector of industry, and the most powerful driving force of growth for material production processes, due to the innovations in global digital electronic telecommunications and information networking and modeling, then we can more clearly see how, according to Félix Guattari, "(...) capitalism takes hold of human beings from the inside," and, "[i]ndividuals are 'equipped' with modes of perception or of standardized desire, for the same reason as factories, schools, and territories."²² In the drive for ever-greater productive efficiency, by switching the relations of production to place the subjectivities of the constituent parts of the market as its principal product, late-stage capitalist industrial production has had to venture into, and interweave itself directly onto, the perceptual, affective, sensorial, linguistic, and cognitive behaviors of people. This interweaving is precisely the "work" (i.e., the production processes *and* processes of distribution) of immaterial labor, but crucially *it is also its product*. Immaterial labor not only transforms the perceptual, affective, sensorial, linguistic, and cognitive behaviors of people, but crucially these behaviors produce (and reproduce) immaterial labor processes and their products. Under post-Fordist labor models, the immaterial labor undertaken is the operation simultaneously of the processes of production, the relations of production, the relations of distribution, *and* the relations of capital. In effect, immaterial labor is the total absorption of capitalist alienation *as* perceptual, affective, sensorial, linguistic, and cognitive behavior during their dynamic modulation by immaterial labor as the enactment of "self-hood" of individuals *and* groups. In this way, the production of "the self," both at the individual and collective level, becomes an industrial concern, and subjectivity a category of commodity form.

Post-Fordist labor models and the immaterial labor that constitutes them are the very structure and fabric of what Gilles Deleuze has famously described as "the societies of control." Differentiating the societies of control from disciplinary societies, as outlined by Michel Foucault, Deleuze shows how the task of the individual within such a society is never-ending and without measure of success. Crucially though, according to Deleuze, the principal difference is that the terms of success in a society of control are constantly shifting and perpetually withheld:

> In the disciplinary societies one was always starting again (from school to the barracks, from the barracks to the factory), while in the societies of control one is never finished with anything – the corporation, the education system, the armed services being metastable states coexisting in one and the same modulation, like a universal system of deformation. (...) The disciplinary societies have two poles: the signature that designates the *individual*, and the number or administrative numeration that indicates his or her position within a *mass*. (...) In the societies of control, on the other hand, what is important is no longer either a signature or a number, but a code: the code is a *password*, while on the other hand the disciplinary societies are regulated by *watchwords*. (...) The numerical language of control is made of codes that mark access to

information, or reject it. We no longer find ourselves dealing with the mass/individual pair. Individuals have become *"dividuals,"* and masses, samples, data, markets, or "banks."[23]

Such pervasive paradigm shifts in the social relations of postindustrial late-stage capitalist societies have profound effects on the relationship that "dividuals" have to time and space as much as they do on their (inter-)relationship with one another, albeit that there are no longer any "ones" nor "anothers" [*sic*] in such a society, as "dividuals" bleed into "oneanother" [*sic*] in a fluid undulating stream of information. The nature of space and time under the sign of such societies becomes increasingly malleable, mutable, and indistinguishable, adding to the fluidity of the relations of time and space and subjects and subjectivities:

> The different internments or spaces of enclosure through which the individual [in the disciplinary societies] passes are independent variables: each time one is supposed to start from zero, and although a common language for all these places exists, it is *analogical*. On the other hand, the different control mechanisms [in the societies of control] are inseparable variations, forming a system of variable geometry the language of which is numerical [i.e. digital] (which doesn't necessarily mean binary). Enclosures are *molds*, distinct castings, but controls are a *modulation*, like a self-deforming cast that will continuously change from one moment to the other, or like a sieve whose mesh will transmute from point to point.[24]

Where once the factory, the school, the home, the barracks, the church, or even the shopping mall were the distinct spatial locations where distinct forms of disciplined and disciplining behavior took place at distinct times for distinct periods of time, which did not overlap, now all elements of life and work are, potentially at least, seamlessly imbricated into one fluid undulating field of *indistinct* forms of controlled and controlling behavior *as signs* (i.e., as information, which also operate with the characteristics and power of *capital*). In such a society, the relations of production, the relations of distribution *and* whatever is left of social relations become asymptotically symbiotic, their changing range of values and coordinates constantly shifting according to dynamic rules that are governed parametrically and that *can* only be discerned and described parametrically, if at all.

That much contemporary architecture today is so beholden to parametrics as its most powerful form of calculation is no surprise, but it is also not in itself a problem per se. The problem that we face is that architecture that deploys parametrics as its principal form of calculation usually does so in order to maximize efficiency for values that are in fact outdated hangovers from now outmoded models of society (the disciplinary societies of the eighteenth-, nineteenth-, and early twentieth centuries, or before them the societies of sovereignty[25] of the medieval and feudal periods of the preceding centuries); namely monetary "profit." The search for ever-greater financial efficiency (i.e., profit-making potential) is testament enough to this point: money being both a disciplinary tool (as a direct enumeration of power potential) and a marker of sovereignty (the quasi-divine sovereign right of the individual to own private property). The confusions and contradictions that arise in the postindustrial late-stage capitalist societies do so largely because they are still in the slow process of moving into the full-blown mode of societies of control.

Where we see architects attempting to design modular, multiscalar, reactive and interdependent constructions that are up to the task of the new social modes of work and communication, we so often hear the well-meaning rhetoric of "flexibility" and "efficiency" in descriptions of the buildings, and in the processes that are entailed in their construction and their usage. However, within the shifting parameters and dissolution of spaces and times of work (within post-Fordist labor models) and the obscure dynamic codes of responsibility (and the diffuse forms of self-controlling within post-Taylorist management models) that such buildings aim to respond to, emulate and propagate, what is often misunderstood is that "flexibility" is not the same as *freedom* and "efficiency" is not the same as *productivity*.

However, we can propose that parametric mathematics is the most suitable mode of calculation to administer, analyze, order, and ultimately fashion such societies of control because its parallel algebraic operations are able to map the convergences and divergences in the minutiae of sometimes opposing and contradictory principles or rules that govern the behaviors of the subjects that constitute such societies. Such societies have what we can call a parametric ideology: a quasi-meta-ideology that eschews all the characteristics of other ideologies *and* ideology in general. The character of the parametric ideology is one for which the dogmatisms, disciplines, principles, coherence, systematicity, and progressions (i.e., the monumentality and enclosures) inherent to the structure of ideologies in general are deformed and dissolved into a complex of multiple undulating "polymentalities" where the only meta-principle in operation is that of transgressing all limit conditions and horizons of principles per se, enacting their deformation.

Parametrics as a mathematical operation is able to construct a malleable net of coordinates that can switch allegiance to and governance by such rules from one set to the another according to how the inputted data changes and/or it can obey both at once without incurring the constriction of contradiction. In this way, parametrics is the most suitable mode of calculation to preside over the massive streams of information that are gathered about the behaviors of subjects in such societies because via parametrics it is possible to make sense of information that comes from behaviors produced by (simultaneously (re)producing) powerful contradictions.

This is how parametrics, in general, acquires its political potential in an age of doubly contradictory social conditions. Because of its massive potential to resolve and/or dissolve such compound contradictions, it is also where it finds its political agency.

Parametricist architecture then is the most suitable and appropriate form of constructing (simultaneously) both the built and electronically networked environment for such a society because such a society no longer requires monuments (if only to deform and dissolve them as societies of control deform and dissolve the ideologies that they symbolize) and/or, as Deleuze points out, no longer requires enclosures (if only to deform and dissolve them as societies of control deform and dissolve the disciplines that they enforce). What such societies need is buildings that are both figures *and* grounds that are conduits through which the "dividuated" parts of the "dividuals" can flow, imbricate, migrate, and mutate as unencumbered as possible. Transgression is the key to this topology as both spaces, times, forms, and identities transgress their own horizons and limit conditions; flowing, migrating, imbricating, deforming, and mutating, simultaneously converging and diverging, amassing and dispersing interchangeably as information, disclosing, dissolving, and sublating the shapes and forms of the disciplines that give way to the forces driving the acceleration of their pace.

If, as has been suggested by immaterial labor theorist Maurizio Lazzarato,[26] work has become a form of art, and *vice versa*, then the new parametricist architectures may well become the endless theaters, dance halls, screens, and instruments where whatever "we" have become *and* where our "dividuated" atomized avatars enact "our/their" lives as a totally operative production of symphonies of flowing information. It will be those who can inter-iterate and cascade their "dividual" selves who will become the composers of these endless cyclonic intersecting self-*gesamtkunstwerks*, and they will be the truly political agents of the coming age of parametric work, parametric space-time, and parametric architecture.

Under such conditions, new forms of work will embody and enact the aesthetics of information and will be governed and driven by the harmonics of data flows, and the work of the architect will be to prepare the score.

Notes

1. Post-Fordist labor models are *modes* of production that occur outside of spatiotemporally constrained locations. They are also *models* of "work" that can occur during traditionally nonwork activities (i.e., leisure, or consuming). The principal products of post-Fordist labor are immaterial and include signs, languages, codes, and subjectivity. By contrast, Fordist labor models occur at specific times and in specific places, involving highly refined divisions of labor. Fordist labor models principally aim to produce material products or services that modulate matter, whereas post-Fordist labor models principally produce information and experiences leading to the modulation of subjectivities.
2. Taylorism is a system of scientific management invented by Frederick Winslow Taylor, outlined in his 1911 book *Principles of Scientific Management*. Taylorism atomizes production processes determining which movements of workers are strictly necessary. A vertical pyramidal hierarchy of a few skilled managers and technicians oversee a much larger number of semiskilled and unskilled workers whose jobs are broken down into the simplest of operations. By extreme contrast, post-Taylorist management employs various forms of dynamic "horizontal" management where any given employee's levels of responsibility can and does change according to the changing needs of the overall production process. Also, the responsibilities of individuals or teams of employees overlap, leading to constant internal competition plus dispersal and obscuring of the structure of power therein.
3. "Relations of production" is a concept developed by Karl Marx and Friedrich Engels describing the total matrix of social relations that a society must engage with in order to continue to reproduce the structure of itself; often referred to as the "socioeconomic structure." The idea is most fully developed in Marx's, *Das Kapital, Kritik der Politischen Ökonomie (Capital: Critique of Political Economy) Volume I*, first published by Verlag von Otto Meisner in 1867.
4. "Relations of distribution" is a more complicated concept than "relations of production" because it is an intrinsic part of the production process overall and its relations of production. Marx outlines this in his unfinished 1858 *Grundrisse* (first published in 1939, by Verlag für Fremdsprachige Literatur, in Moscow), *and in Chapter 51 of Das Kapital, he explains that the total "socioeconomic matrix"* includes production, circulation, distribution, and consumption:

   > [i]n society (…) the producer's relation to the product (…) is an external one, and its return to the subject depends on his relations to other individuals. (…) Distribution steps between the producers and the products, hence between production and consumption, to determine in accordance with social laws what the producer's share will be in the world of products. (…) Distribution is itself a product of production, not only in its object, in that only the results of production can be distributed, but also in its form, in that the specific kind of participation in production determines the specific forms of distribution, i.e. the pattern of participation in distribution. Ibid.

5. Jean-François Lyotard, Chapter 1, "The Field: Knowledge in Computerized Societies," in *The Post Modern Condition: A Report on Knowledge*, trans. Geoff Bennington & Brian Massumi (Manchester: Manchester University Press, 1984), p. 4. Originally published in France in 1979 by Les Editions de Minuit.
6. Ibid., p. 5.
7. Patrik Schumacher, "Parametricism as Style—Parametricist Manifesto," London 2008. Found at http://www.patrikschumacher.com/Texts/Parametricism%20as%20Style.htm [accessed July 17, 2014].
8. Ibid.
9. Ibid.
10. Ibid.
11. Quoted in: John Ruskin, Sir Edward Tyas Cook, & Alexander Wedderburn Dundas Ogilvy, *Biblioteca Pastorum: The Economist of Xenophon*, Chapter IV (London: G. Allen Publisher, 1907), p. 47.
12. Adam Smith, 1776, *An Inquiry into the Nature and Causes of the Wealth of Nations*, Book I, Chapter I, "Of the Causes of Improvement in the productive Powers of Labour, and of the Order according to which its Produce is naturally distributed among the different Ranks of the People." Found at http://www.econlib.org/library/Smith/smWN1.html#B.I,%20Ch.1,%20Of%20the%20Division%20of%20Labor [accessed July 17, 2014].
13. Charles Babbage, Chapter 19 "On the Division of Labour," in *On the Economy of Machinery and Manufactures*, 1832, line 217. Found at http://www.gutenberg.org/cache/epub/4238/pg4238.html [accessed July 27, 2014].
14. Karl Marx, "First Manuscript," in *Economic and Philosophical Manuscripts*, 1844; found in T. B. Bottomore, *Karl Marx Early Writings* (London: C.A. Watts and Co. Ltd., 1963), p. 72.
15. Here, pricing of goods in a capitalist market is referred to as "ineffable," using both the sense that the calculation of the production process, the relations of production and the relations of distribution, that lead to any given pricing are so complex as to be practically incalculable and therefore unthinkable, *and* that such a condition renders the "price as index of social relations" as a *quasi-sacred* value and hence is always and already ideological.
16. Karl Marx, *Capital, Vol. 1* (Moscow: Progress Publishers, 1887), p. 333.
17. Friedrick Engels, "The Part Played by Labour in the Transition from Ape to Man," written in May/June 1876 (first published in *Die Neue Zeit* in 1895).
18. Karl Marx, Book I of the *Critique of the Gotha Program* (written in April 1875 and first published in an abridged version in *Die Neue Zeit*, Vol. 1, No. 18, in 1890).
19. "Lean Production," or "Toyotism," is a management ethos that emerged throughout the 1980s and is closely associated with the Toyota Productions Systems Company. It is a system for manufacturing that eliminates all areas of waste in the production process at all levels, adding maximum value for customers, and simultaneously increases profits for the company. For more on this, see Mattias Holweg, "The Genealogy of

Lean Production," *Journal of Operations Management* Vol. 25, No. 2 (Philadelphia, PA: Elsevier Press, 2007), pp. 420–37.

20. Exemplary of this is the Nike Corporation, which disperses production to independent companies worldwide, each competing to design, manufacture and/or assemble, distribute, manage, and administrate the component parts of the sportswear. Nike itself is unencumbered to work on the creation of "images"; both the constant recreation of the image of the brand and simultaneously and symbiotically the image(s) of its current and future customers' lifestyles and desires (i.e., the subjectivity of the subjects of Nike's "image-making"). For more on this, see Robert Goldman & Stephen Papson, *Nike Culture: The Sign of the Swoosh* (London: Sage Publications, 1998).
21. Maurizio Lazzarato, "General Intellect: Towards an Inquiry into Immaterial Labour," in *Immaterial Labour: Mass Intellectuality, New Constitution, Post-Fordism, and All That* (London: Red Notes, 1994), pp. 1–14.
22. Félix Guattari, *Molecular Revolution: Psychiatry and Politics*, trans. Rosemary Sheed (Harmondsworth: Penguin, 1984).
23. Gilles Deleuze, *Postscript on the Societies of Control*, Vol. 59 (Cambridge, MA: MIT Press, 1992), pp. 3–7.
24. Ibid., p. 4.
25. "Foucault has brilliantly analysed the ideal project of these environments of enclosure, particularly visible within the factory: to concentrate; to distribute in space; to order in time; to compose a productive force within the dimension of space-time whose effect will be greater than the sum of its component forces. But what Foucault recognized as well was the transience of this model: it succeeded that of the *societies of sovereignty*, the goal and functions of which were something quite different (to tax rather than to organize production, to rule on death rather than to administer life) (…)," ibid., p. 3.
26. "(…) globalization has drastically altered the concepts of art and work through two parallel and complementary trends. The first assimilates artistic and all cultural practices into the capitalist economy by transforming artists into 'workers,' and 'citizens' into various publics. The second turns 'creation' into a process that traverses art, the economy, science, and the social sphere, by transferring the problems, methods, and practices that were thought to be specific to art into other domains." Maurizio Lazzarato, "Art and Work," *Parachute 122 "Travail ** Work"*, June, 2006. Paragraph 1. Found at http://www.parachute.ca/public/+100/122.htm [accessed July 1, 2014].

Chapter 9

Undelete: Recreating censored archives

Laura Kurgan and Dan Taeyoung

Use of the internet in China is policed—watched over, censored, and punished—by a concerted government effort that has been nicknamed "The Great Firewall." The aim is to keep politically unacceptable or "sensitive" content invisible to Chinese internet users—for example, discussions of the Tiananmen Square massacre. Twitter and Facebook are largely blocked, as are many news outlets and human rights web sites. Web searches are seriously curtailed, politically charged words are restricted, and online postings and other content are routinely removed from the feeds and pages of those who could, potentially, create a movement inside the country. It is a system of control. For many Chinese users who wish to access blocked web sites, the only option is a Virtual Private Network (VPN). The word *private* reminds us that, in fact, the internet constitutes a form of public space in China, just as elsewhere, in which people gather, connect, exchange, and communicate—for all sorts of purposes, including political ones.

Our project, Jumping the Great Firewall, examines some innovative strategies employed by users of Sina Weibo, a Twitter-like micro-blogging platform, to evade and delay censorship. Sina Weibo, like Twitter, allows users to post messages of up 140 characters, as well as to directly attach images to their posts. Textual censorship on the Chinese internet is automated with algorithms: text is relatively easy to analyze on a massive scale with computation, and so the removal of language deemed dangerous is quick and efficient. Images are much more difficult to analyze computationally, and so their content is filtered by human censors. Chinese Weibo users have understood this technical disparity between word and image, and have exploited it in powerful and creative ways. When writing about a potentially controversial topic, users can post their messages as screenshots of text or as photographs of handwritten notes, effectively preventing them from being filtered algorithmically. Handwriting, for instance, stymies even the best of optical character recognition software. While these posts are eventually seen and removed by human eyes, the delay gives messages and ideas time to be seen by others, reproduced, and spread.

Our project started with the aim of recovering and recreating an archive of these censored images. The archive demonstrates both the restrictive nature of Chinese government policy and the proactive creativity of Chinese citizens and activists. It also restores and makes visible once again what had been deleted, allowing others to participate in ongoing debates and to understand and measure the extent and the focal points of censorship. What were the taboo topics—and what were people saying about them? How did the messages spread—how were they copied, reposted, and multiplied?

We began by picking a discrete sample of people—100 civic activists, mostly journalists or lawyers, each of whom had over 2,000 followers. We selected our group from a much larger pool of people who were already being followed by WeiboScope, a project of the University of Hong Kong that, since 2011, has been archiving the deleted Weibo content of some 350,000 people with more than 1,000 followers.[1] We gathered our data from this group of 100 independently, and structured the collection protocol so that we not only archived the deleted image-posts, but also tracked how many times they were reposted and how long the posts survived. Our project is meant to complement theirs with a searchable, "browsable," and visual data set, telling a more detailed story about a small part of the archive at a specific moment in time.

To generate this archive, we wrote a series of Python scripts that accessed Weibo's Application Protocol Interface (API).[2] Like Twitter, Sina Weibo has an open API, which enabled the automated collection of posts between July 24 and August 4, 2013, and between September 8 and November 13, 2013. Every minute, image-posts by our 100 users were captured and stored in a database. We wrote a script to revisit each of these captured posts on Weibo once a minute in order to verify its continued existence in its original context. Weibo's API is not completely open; among other constraints (such as not being able to look up information about users), it narrowly limits the number of hourly visits per user. To overcome this problem, we created twenty-five puppet user accounts, which enabled us to track all the image-posts of our 100 posters every minute. In exchange, the scripts were only able to track posts that were less than four hours old. As such, we made a concerted effort to trace the short-term, rapid-fire censorship of these image-posts, especially as these posts constituted a large percentage of total censored posts.[3]

Undelete: Recreating censored archives

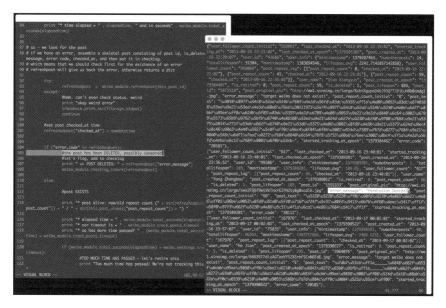

Figure 9.1 Python script detail, Jumping the Great Firewall, Advanced Data Visualization Project, GSAPP, Columbia University.

Over the course of the collection period, we tracked and checked for a post over ten million times. Each time, we verified the existence of the posts of interest, and noted whether they were still in their original location, had disappeared, or had been reposted elsewhere. Every once in a while an individual post would suddenly return a specific error message upon our revisit—*20112: Permission Denied*. This told us that the post had not been removed by its author, but had been censored.[4] The script would then log the date and time of the post's disappearance, retrieve records of its last sighting, and store and log the event in the historical archive for future reference.

In the September to November collection period, out of 34,837 Weibo posts containing images from the 100 activists, 952 posts were deleted, and of them 334 posts (0.96 percent) were conclusively identified as having been censored.[5] On an average day, 528 image-posts were created and 5 were deleted, and on average, one post was deleted approximately every four hours and forty-five minutes. Considering that these image-posts were removed by nonautomated, human censors, the deleted posts had a surprisingly short lifespan. Over a quarter (27.2 percent) of the deleted posts were removed within four minutes of posting, and more than half (50.5 percent) were deleted within twenty-four minutes of posting. In a few cases, posts were deleted nearly immediately—12 percent of the deleted posts disappeared within the first minute.

To display this phenomenon, we created a visual interface for examining the archive of images that we had collected. We wanted to represent each deleted post as an active event of censorship, rather than simply a record documenting the deletion of a post in a historical past, which is what WeiboScope does very well. We assembled the deleted posts on a timeline with symbols that represent both lifespan and repost count, which is to say, the posts' growing popularity. A short, tall triangle indicates that the posted image was reposted many times and deleted in a short amount of time. A more elongated, narrow wedge shows that the posts took a longer time to disappear, perhaps because they had been reposted less frequently and hence escaped notice longer. Posts are isolatable by users, allowing a viewer to scroll through the graph and find out that, for example, some of the most dynamic images that spread and were removed most quickly were posted by the same person.

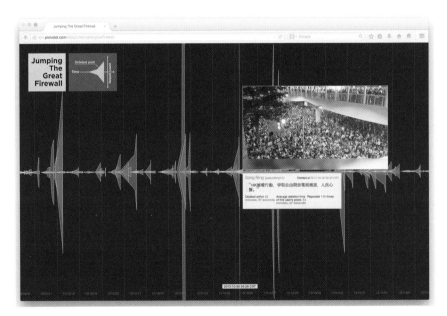

Figure 9.2 Visual interface, Jumping the Great Firewall, Advanced Data Visualization Project, GSAPP, Columbia University.

To protect users' identities, the data archive itself is anonymized for public display. All images are blurred to obscure identifying captions, and all posting dates and times are offset by random increments so that a post could not be easily located by their specific timestamp. Finally, usernames are replaced with pseudonyms.

Using the two-week July to August database, which we collected in collaboration with the online investigative data-journalist team at ProPublica,[6] we translated a sample of 500 images and found that the topics of the image-messages could be organized into ten general categories: Political Speech, Long Text, Dissidents, Bo Xilai, Public Figures, Protests, Wang Lin, Scandals and Corruption, Censorship, and Humor and Cartoons.[7] In those two weeks, we collected a total of 7,972 posts, of which 1,710 had been deleted. Of the deleted posts, we identified at least 557 posts that had been removed by censors.[8] In the deleted postings, the most prevalent categories of censored posting were "political speech" and the dramatic-looking "*longweibo* texts," i.e., postings that exceed the 140 character limit. In a two-week sample from the 100 people we tracked, we found that 24 percent or 136 posts were *longweibo*s. Ten percent of the posts were about Bo Xilai, a Chinese politician publicly accused of corruption and sentenced to life imprisonment—the story about Bo broke during this two-week period.

Search

It appears self-evident now, but a central component of the excitement around Gmail's introduction in 2004 was the presence of Google-caliber *search* within your email, aided by its (at the time) staggeringly large amount of storage space. Gmail's initial tagline proclaimed: "Gmail is an experiment in a new kind of webmail, built on the idea that you should never have to delete mail and you should always be able to find the message you want." As the "archive" button replaced "delete," searching, not sorting, was championed as the primary cognitive action for the data of the future. Ten years into that future, search functionality is expected and resides in all forms of media: web sites, phones, blogs, apps, even cable TV set-up boxes. It is understood that any web site is a repository of data, and that the data can be aggregated, sorted, searched, and displayed. Often, the search is pre-executed for the user: "Other films you may like," "Other consumers bought," "Based on your search, you may like." Specialized search engines are proliferating, searching within electronic components (Octopart), startup jobs (AngelList), real estate listings (Zillow, Trulia). Search itself is a product, provided as a form of infrastructure to other software developers: "infrastructure-as-a-service."

From a technological perspective, searches or feeds (as distinct from the data archive) provide an interface. From a user perspective, these interfaces generate an aesthetic of visibility. When Google informs a user that she is viewing the first ten of "about 930,000,000 results," search borders on the absurd, becomes theatrical—it is impossible to see or make sense of the search result as a list. Retrieving 930 million results from a search term is a question of scale and not just of size, of uncertainty rather than certainty. The search engine tells you: all data is available, at an incomprehensible scale. It just has to be searched.

By contrast, encountering Twitter or Weibo feeds is not about searching, but more akin to wading through Heraclitus's ever-changing river, engaging an interface with the promise of never-ending, real-time "raw" information. Like its fluvial counterpart, the stream changes when information is updated. Immediacy (or in tech-terms, "low latency") is the key behind Twitter's popular use as a news stream during disasters, protests, or emergencies. If search creates an expansive space, then feeds render movement, a deluge of a hyperactive, seemingly unfiltered present.

Posts on Weibo generate both a searchable archive and a feed, but they look very different. The visibilities generated through search engines and feeds are fabricated at all levels.[9] When you post on Weibo, you might be censored. Taking an activist approach to this fact, we fabricated an archive and search engine specifically for the Weibo feed, obsessively checking back in on our user sample so that we could recover posts that had been deleted by the censors. Weibo cloaks censorship under the aesthetics of the real-time feed—following the feed, without exploring to the archive, most of its users have no idea what is being censored and what is not.

For our project, distrusting the appearance of the real-time feed meant creating our own historical archive through an automated and repetitive checking process. The importance of this frequency was made visible through the collection process itself. At one point, we altered the scripts to check every hour, then every minute from the feeds. And as we did, we discovered a large number of posts we had previously been blind to. Similarly, trawling Weibo for new posts only once in twenty-four hours means that that the crawling tool, and therefore the archive, ignored and hence omitted collected posts younger than twenty-four hours. Collected this way, the vast majority of posts would have already been posted, scanned, censored, and removed within this timeframe. The fluid stream of posts from users would have appeared and disappeared by the time the scripts would have accessed them. It was only when the scripts began to check every minute (or less) that the nearly real-time deletions become visible within the archive.

Other activist approaches to censorship (or its sibling, surveillance) include the "warrant canary," a method not unlike our collection process. The warrant canary is a sign or method to communicate the presence of a secret subpoena through the absence of a publicly displayed message. Popularized by librarian

Jessamyn West in 2002 in response to the Patriot Act's demand for libraries to secretly provide lending records. An example "canary" may be a sign reading: "The FBI has not been here; watch closely for the removal of this sign."[10] The future absence of the sign, of course, would indicate the past presence of a subpoena, a fact that under the law could not be stated in the affirmative. The warrant canary relies on the ever-vigilant sharp-eyed observer who notices the blank space on the wall, and who is able to compare his or her memory of a sign to its present absence. In the same way, our Weibo crawling scripts had to amass a secondary database, archiving disappearances by checking the present as frequently as possible against the "past-present."

Our methodology for the examination of censorship in Sina Weibo produced a visceral understanding of how easily data presented as a feed or a stream is able to mask deletions. On the one hand, on Weibo, real-time flows present a continuously censored narrative. On the other hand, the explicit censorship and removal of posts meant that we could create an archive of censorship. If searching or streaming involve exploring an archive with two different modes of experience, either with the aesthetic of scale or that of unfiltered flows, then our reconstituted archive operates at the level of a small, static sample, as a proof-of-concept of visibility. We verified that the practice of image-posting is rapidly censored, but also that the images spread rapidly before they are deleted entirely. The spread of images can only frustrate the human censors—it is likely impossible to find all of the reposts.

In collecting and visualizing this archive, we created the possibility of mapping the topics and flow of dissident communication, and also highlighted both the censorship regime and the efforts by activists to circumvent it. Because of the architecture of the open API, even with its restrictions, censorship does not simply make information disappear. It leaves traces and produces effects, and can itself become a source of information.

The project exists as a prototype against censorship, not as a viable solution: scaling this project up would be a massive, probably impossible undertaking. No one except the Chinese censors themselves, and their algorithms, could produce a complete archive of censorship. Rather, it is our hope that this small project provides a window into what is being censored and the recovery of what has been deleted; as well as providing another piece of evidence that the architecture of the internet ensures the impossibility of total censorship. This is not to say that the practice of censorship is not an effective one. But there are creative ways that people in China are protesting and succeeding in making their voices heard by others. Creating and visualizing the archive itself is our small gesture of solidarity with those voices.

Notes

1. WeiboScope data can be downloaded at http://weiboscope.jmsc.hku.hk/datazip/ [accessed November 20, 2014]. For more details on our methodology in collaboration with ProPublica, see http://www.propublica.org/article/how-we-observed-censorship-on-sina-weibo [accessed November 20, 2014].
2. Python is a powerful, high-level programming language used for many applications including web applications, statistics, machine learning, artificial intelligence, and general-purpose scripting. In this context, Python was used as (1) the language highly adept at handling Unicode strings and data, and (2) the nature of the collection process required that the scripts run repeatedly every minute (via cron) on a Linux server.
3. We learned this from Tao Zhu et al. "In our data set, 5% of the deletions happened in the first 8 minutes, and within 30 minutes, almost 30% of the deletions were finished. More than 90% of deletions happened within one day after a post was submitted. This demonstrates why a measurement fidelity on the order of minutes, rather than days, is critical." Tao Zhu et al., "The Velocity of Censorship: High-Fidelity Detection of Microblog Post Deletions," in *Proceedings of the 22Nd USENIX Conference on Security*, SEC'13 (Berkeley: USENIX Association, 2013), pp. 227–40, also found at http://dl.acm.org/citation.cfm?id=2534766.2534786 [accessed November 19, 2014], p. 4.
4. Previous researchers ascertained the error code *"20112: Permission Denied"* to be evidence of censorship. Ibid.
5. In our research, we found another error message (*"20101: target weibo does not exist!"*) accompanying 610 deleted posts that could have either been censored or may have simply been a result of the user deleting their own post. As it was not possible to conclusively determine whether these posts were censored, these posts were not included in our representation.
6. For details on our work with ProPublica, see http://www.propublica.org/article/how-we-observed-censorship-on-sina-weibo [accessed November 20, 2014].
7. ProPublica, "China's Memory Hole: The Images Erased from Sina Weibo," in *China's Memory Hole: The Images Erased from Sina Weibo*, November 14, 2013, found at https://projects.propublica.org/weibo/ [accessed November 20, 2014].
8. ProPublica's process found that 6.98 percent of posts were censored, as opposed to 0.96 percent of posts being censored from our two-month collection of posts. As mentioned above, our database checked for a post deletion every minute (with the trade-off of disregarding posts that were not deleted after four hours), while ProPublica's database was checking every six minutes. While we are not fully certain why the percentages of censored posts may be different, we suspect that our archive was able to capture fewer posts that were, however, more controversial and thus deleted within a shorter time span.
9. "Facts were facts—meaning exact—because they were fabricated—meaning that they emerged out of artificial situations. Every scientist we studied was proud of this

connection between the quality of its construction and the quality of its data." Bruno Latour, *Reassembling the Social: An Introduction to Actor-Network-Theory* (Oxford & New York: Oxford University Press, 2007), pp. 90–91.
10. Jessamyn West, "Five Technically Legal Signs for Your Library," in *Five Technically Legal Signs for Your Library*, December 18, 2002, at http://www.librarian.net/technicality.html [accessed November 17, 2014].

Chapter 10

Disputing calculations in architecture: Notes for a pragmatic reframing of parametricism and architecture

Andrés Jaque

The discussion on parametricism in architecture needs to be reframed. The desire for a new technologically driven march of progress has gained an amplified voice that conceals a much richer ecosystem of heterogeneous realities. With the intention of counterbalancing that hegemony, I propose to look at parametricism, calculation, and digital technologies as constituent components of daily life. By looking at pragmatic accounts of two cases of domestic urban enactments in which calculation, parametricism, and digital technologies are playing key constructive roles, we can gain opportunities to introduce alternative perspectives that reconnect the discussion with the empirical capital that daily life accumulates.

Figure 10.1 shows the enactment composed by the entities participating at a very specific moment in the making of a particular micro-society: an extended family that is a segment of a larger social construction, the Mouride Brotherhood. Part of the extended family lives in a farmhouse in Touba, Senegal. Another part of the family works and lives distributed between two European cities: Madrid and Paris. The economy and the welfare of the micro-society are based on the possibility of keeping those family members who are female, elderly, disabled, or very young in Touba, where their living expenses can be minimized, while having the young males in wealthier cities of European countries, where they maximize their incomes by selling counterfeit products such as DVDs and fake Louis Vuitton bags. The image reconstitutes the most important entities that participated in a tiny event of great importance in 2010: the initial preparation for the future displacement of a male Touba-based teenage member of the family,

Disputing calculations in architecture 169

Figure 10.1 Reconstruction of the urbanism of a Mouride extended family distributed between Paris, Madrid, and Touba. Andrés Jaque/Office for Political Innovation 2012.

in advance of his reinstallation in the Lavapiés district in Madrid: a process by which he would become an active contributor to the economy of the whole and by which he would gain grown-up status within the group. As part of a five-year project that the Office for Political Innovation has developed, to systematically study more than 100 cases of *ordinary urbanisms*, we have traced the relational extensions that happened in this particular moment.[1]

Lavapiés is composed of both long established social groups and recently arrived immigrants. This combination is largely due to the effect of an extensive fragmentation in the ownership structure of the buildings in the neighborhood and the vast social transformations that the city of Madrid has experienced in recent decades. Wholesale facilities, tourist-oriented boutique hotels, refurbished apartment buildings, and a large number of deteriorating dwellings can now all be found within a short walking distance of each other. Since access by Senegalese nationals to countries like Spain and France is severely restricted, young Mouride males tend to arrive in Madrid or Paris as illegal immigrants. The subway is the place where the police most easily seize undocumented migrants. Lavapiés offers the possibility of accommodating all the activities that structure the life of the young Mouride males within easy walking distance: wholesale facilities (where ready-to-sell products such as DVDs or fake Louis Vuitton bags are available), tourist sites, and inexpensive deteriorated apartments (where illegal immigrants are likely to be accepted). Living in Lavapiés helps the Mouride males avoid the risk of police detention that the subway contains. The preparation sequence was sensitive as well to the need to solve language and orientation difficulties for the new immigrant upon arrival in Madrid or Paris. The presence of African grocery stores and Senegalese restaurants is vital to empowering the Mouride community in Madrid, for it provides an entry to a public space where the knowledge and capacities of the Senegalese individuals are recognized and performed.

The sequence started the moment that one of the young boys living in the family farm in Touba decided to emigrate to Madrid. At this point, the family matriarch at the farmhouse in Touba called one of the males living in Madrid, an older cousin of the boy. The cousin did not answer his cell phone but, instead, headed to a phone parlor where he could obtain better rates for international calls. There, he called the family matriarch to learn of the next arrival. He asked her to stress the need for the boy to walk all the way to either an African grocery store or a Senegalese restaurant in the Lavapiés district. The plan succeeded and, several months later, the young boy made his way to the African grocery in Lavapiés where he found people who put him in touch with his relative. He then took a place in a shared apartment with his cousin and four other Mouride men.

It is important to consider the nature of the urban composition in which this event developed: not a city but a fragmented transnational assemblage that can be best explained using Actor Network Theory.[2] In this urban constellation, built devices—such as the apartments, mosque, phone parlors, African grocery stores, and Senegalese restaurants in Lavapiés—are activated in the urban scene only by interacting with a number of diverse technologies including cell phones, rugs, speakers, online platforms, and money transfer services. This urbanism is not shaped by the city itself—its grids and the volumes and spaces of its buildings—but by an association of heterogeneous devices that interact to produce an ecosystem of heterogeneous entities. Fragments of this constellation can be found in shared spaces collectively constructed in the minds and books of the Mouride believers. These fragments are connected by interaction and the performativity of urban dynamics. They gain continuity when phone calls are made, money transfers are ordered, and the relatives of recent immigrants are informed of arrivals. The urbanism by which the Mouride family is enacted is not fixed but performative.

Such an urbanism challenges the ways we think politics is embodied in architecture. In recent years, this issue has compelled a number of theorists and practitioners to align themselves with one of two positions: *techno-determinism* or *techno-neutrality*. The determinists argue that the form of the city and its architectural conditions cause societies to emerge in the ways they do. The neutralists, however, believe that architecture is a neutral actor that can potentially contain *any* social form. In Figure 10.1, neither of these alternatives can be applied.[3] There is not a single architectural device within the image that could alone produce the society depicted. The apartment where the six Mouride people lived could not create on its own the urbanism of everyday life performed in its interior. A vast range of devices collectively builds this fragmented-but-interacting urbanism, although it is also true that the designs of the individual architectures are not without agency. The dimensions of the apartment and its position on the street, for example, play significant roles in this particular urbanism. One of the spaces included in the image was once a domestic unit, but at the time of our study functioned as a Mouride mosque. Conditions that catered to its evolved state made its reprogramming possible. The space was diaphanous, and its entrance did not disturb the tranquility of the main room. To become part of a Mouride urbanism, however, the apartment needed to engage "new technologies." It became part of such a dynamic urbanism by housing books and minds inscribed with shared beliefs. It was transformed by the existence of hi-fi speakers and tapestries depicting holy sites. The political agency of architectural devices is shared. The potentials and limitations of each device interact with other entities and together construct a new form of agency.

Politics and construction are embodied not in individual technologies or architectural devices, but in the interaction of their particular potentialities.[4] When considering the intelligences that shape the interactions between different entities, the role played by calculation stands as a momentous factor, for it is through calculation that the members of the extended family are distributed into a discontinuous transnational accommodation. It is calculation of risk that demarcates the activities of the young displaced males within Lavapiés. It is calculation of cost that shapes the composition of technologies and their sequential mobilization in the telephonic communications between the family matriarch in Touba and the cousin in Madrid. In these calculations, socio-geographies are coded in parameters of cost, earnings, duration and populations; and great numbers of alternative versions of these socio-geographies can be discussed and explored. Each calculation was different. In the use of phone technologies, a significant amount of time-demanding work was invested in comparing existing telephone fees. Almost all the members of the family living in Madrid took part in these comparisons by consulting a broad variety of web sites where information on telephone fees was provided in diverse formats. Comparison of this information required it to be recomposed. Data coming from different sources needed to be translated into common parameters to become comparable. The results were then discussed in informal conversations that engaged the whole community in an intermittent conversation that helped create a collective criteria that was applied in the specific telephonic conversation that I have referred to. Even though the principal aim of the collective endeavor of calculation was to reduce the operational cost of the micro-society—something that is obviously a central part of the constitution that keeps it together—other purposes were constructed by the collective calculation process. Competition between individuals to gain authority and prestige through the discussion, or to exclude themselves from the risk it embodied, made social distinctions between the roles that the individuals would play within the group. Updates to the calculations over time provided opportunities for those in the group less adapted to technologies and the practicalities of functioning in the city to be introduced to this knowledge by the more savvy ones. Calculation was in itself a collective activity that contributed to the evolution of the group.[5]

It is important to consider as well that the members of the extended Mouride family were not the only ones making calculations. For instance, the presence of the police in the subway was the result of a spatial calculation.[6] The linearity and unidirectionality of the subway routes, their capacity to concentrate people at peak hours, and the existence of security staff in most metro entrances made it possible for police agents to maximize their capacity to control a great amount of people while minimizing their investment in failed attempts at detention. The fees of the telephone companies, just to provide another example, are the result of complex calculations meant to maximize the exploitation of their investment in infrastructural resources. Even though the calculations that the telephone companies and police engage in are not included in the image, they are an essential part of the enactment. The design of the enactment—the way the different entities that participate in it are composed by the performance they all participate in—is a direct reaction to these other calculations and to the techno-social designs they produce.

In Figure 10.2, a single mother and her ten-year-old son live on the outskirts of the city in a rented apartment on the same block as the mother's parents, who can take care of the son while the mother works. Such a relational scheme shapes the way the mother emerges as a component of urban life. This urbanism allows her to use social media sites such as Match.com, where she develops romantic relationships, and to use her parents' apartment in Madrid's city center, where her online sexual relationships become offline sexual encounters. Even though she would have preferred to live in the city center, by living close to her parents, she reduces her need for babysitting, reduces her living expenses, and still keeps her career active and competitive. The mother made her decision after a deliberative process in which she discussed alternative scenarios and the economical schemes they activated with a number of friends and relatives. That process could be explained as a prolonged-in-time collective calculation in which each imagined scenario would be reconstructed as a combination of values associated to comparable parameters. Her decision to move to her parents' block did not facilitate the development of an active and changing sexual social life, a sexual life she could not easily accommodate in a residential location on the outskirts of the city, far from nightlife, bereft of spontaneous opportunities for her to find potential lovers. The use of digital technologies here plays a crucial role. Digital technologies produce a space of negotiation in which her daily urbanism could expand and gain multiplicity. When considered in detail, the profiling and negotiation processes leading to potential sexual encounters happening in the online space of Match.com could be described as processes of collective calculation. Profiles reconstruct humans as collections of parametrical options. Age, height, income, distance, urgency, and availability for sex or romance are parametricized as decisive information to be considered when selecting and negotiating potential partners. Digital mediation however does not provide an automatic outcome, but a frame for a tentative trial-and-error

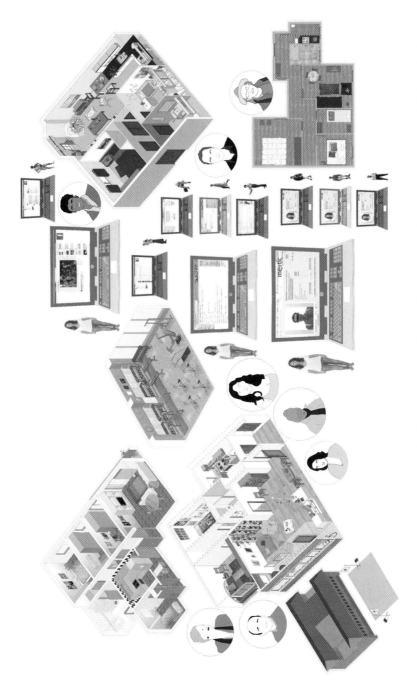

Figure 10.2 Reconstruction of the urbanism of a family group composed by a mother, her son, and a number of humans and devices distributed between London and Valdemoro. Andrés Jaque/Office for Political Innovation 2012.

dynamic, likely to accumulate conflict, politics, failure, and accident. This second case shows how, when performed in real life, calculations are interdependent. Cost reduction required the mother to engage in online calculation for the political project of engaging with others in sexual interaction. Digital technologies in this case did not make analog calculation obsolete—for example, when the mother discussed her options with friends. Both technologies were assembled in this urban enactment and they had a shared performance.

These two cases, based on accounts of the way calculations are performed in ordinary life as a constitutive part of design processes, are relevant for reframing the way notions of parametricism have been operating in architectural discourse in the last few years. They are relevant because they provide a direct opportunity to confront what parametricism is believed to be in certain architectural design circles, where it has become a sign of identity. Here, it is important to make a clear distinction between the rich and varied traditions exploring the effect and potential of digital technologies to accommodate architectural practices, and the very specific and reduced segment of them that have defined themselves as Parametricist. Both the case of the single mother and that of the Mouride group contain designed architectures, such as the mosque, apartments, telephone parlors, and family farm in Touba. These spaces are not just architectures containing the intentions of their first designers, but ones in which calculative social endeavor shapes trajectories by which they became urban enactments. These two cases are made by parametric calculation, but whereas it is most common to find arguments presenting parametricism in architecture as both a new archi-social paradigm and as a "new" modernism, in these two accounts there is no way to define a breakthrough between preexisting pre-parametric enactments and parametricized ones. The possibilities and the potential these enactments make available are based neither on the autonomy of new digital speeds and bandwidth for calculation, nor on their discontinuity with obsolete pre-digital calculation modes. These enactments are instead compositions of diverse modes of calculation, based on diverse technologies, times, and demarcations. Digital calculation, in these cases, is not an advanced version of previous modes of calculation, but one among a number of diverse technological and epistemological regimes.[7] The persistence of these diverse regimes is not a residual part of an evolution process that will lead to the extinction of nondigital technologies. The social relevance of digital technologies is produced by their interaction with resilient nondigital technologies.

Secondly, in the context of parametric design, the outcome of parametricism tends to be considered as the production of form. By form, what is meant in most cases is *fixed material volume at-a-building-like-scale*. In these two cases, in which a complex and maintained-in-time number of parametric calculative forces participate in the making of specific urban enactments, it would be quite inaccurate to say that fixed volumetric form alone is what they *produce*. To start with, the term *produce* could much better be replaced by *mobilized*, and it is definitively a number of diverse social realities—the relationship of a mother with her son, the demarcation of people within a city environment, the way online and offline spaces negotiate continuities and discontinuities—that are enacted by the effect of the parametric calculations these two archi-societies experience.

Finally: whereas parametricism has been the argument to insistently claim the disconnection of architecture from politics, and even to be the first step toward postpolitical architectural practices, in the account of these specific cases, parametricism is not a space of convergence and does not provide social coherence. Calculation-made enactments are hosted by different and evolving social demarcations, performed in desynchronized time sequences, and cater to diverse and, in many cases, confronted interests, ideologies, sensitivities, stakes, and programs. Parametric calculation, in these cases, is not providing consensus, nor a postpolitical society, but rather an assembly of calculations and parametricisms. Politics are materially embodied in the articulation of confronting calculations. It is precisely heterogeneity that was brought into the Mouride enactment by the intense calculative endeavor that the Mouride family performs daily. Heterogeneity was needed, in the form of a multiple demarcation, as a way to adapt to the lack of welfare opportunities in a purely Touba-located scheme. The digitalization of the space for romance did not bring homogeneity to the mother's life, but actually exposed her to a broader capacity to encounter otherness, which affected the heterogeneity of her family's ecosystem. The increase of calculation capacity brings otherness and politics to the system, not homogeneity or continuity.

Notes

1. "Modes of Living," an interviews-based ethnography of contemporary domesticities developed by the Office for Political Innovation with the support of the Empresa Municipal de la Vivienda de Madrid and the European Union. The case studies mentioned in this text are all part of the "Modes of Living" project. Part of the conclusions of this research were exposed in: Jaque, Andrés, "Urban Enactments," *A+U Architecture and Urbanism*, N. 520, [JAP/ENG] 2014 (Japan).
2. Bruno Latour, "Technology is Society Made Durable," in *A Sociology of Monsters Essays on Power, Technology and Domination*, ed. J. Law, Sociology Review Monograph

38 (London: Routledge, 1991). Also Bruno Latour, *Reassembling the Social: An Introduction on Actor-Network-Theory* (Oxford: Oxford University Press, 2005).

3. A broad argument of the way that participation of material devices has been discussed in the last few years can be found in Noortje Marres's "As if Things Mattered," in *Material Participation: Technology, the Environment and Everyday Publics* (Houndmills: Palgrave Macmillan, 2012).

4. As was already exposed in the foundational work by Harold Garfinkel, *Studies in Ethnomethodology* (Englewood Cliffs, NJ: Prentice-Hall, 1967; Oxford: Polity Press, 1984).

5. The way calculation is understood in this text, as a socially situated activity, follows the way the calculability of goods was examined in Michel Callon and Fabian Muniesa's essay "Peripheral Vision. Economic Markets as Calculative Collective Devices," *Organization Studies* Vol. 26, No. 8 (2005), pp. 1229–50.

6. During the first months of 2010, the Periodico Diagonal journalist Eduardo León carried out an extensive report on the use of the subway by Madrid's police force to detect and detain immigrants without residence permission. These police practices have prompted the reaction of a part of the Lavapiés' population, as has been repeatedly reported in the press: see, for example, Miguel Ángel Medina & F.J. Barroso, "Los vecinos de Lavapiés vuelven a encararse con la policía para evitar una detención," in *El País*, Madrid, December 7, 2011.

7. Both exposed cases (the Mouride distributed domesticity and the single mother's composition) are recent cases of complex material design of the urban of great relevance in the making of social advance. Their capacity to compete and provide an alternative to dominant powers, such as the immigration policies of the United States or the real estate market in Madrid, depends on their use of combined technologies, and it is not weakened by their disregard to some of the foundational technologies of parametricism. This makes them dissidents of some of the notions on which parametricism is grounded: "The current stage of advancement within Parametricism relates as much to the continuous advancement of the attendant computational design technologies as it is due to the designer's realization of the unique formal and organizational opportunities that are afforded. Parametricism can only exist via sophisticated parametric techniques. Finally, computationally advanced design techniques like scripting (in Mel-script or Rhino-script) and parametric modeling (with tools like GC or DP) are becoming a pervasive reality. Today it is impossible to compete within the contemporary avant-garde scene without mastering these techniques." Patrik Schumacher, "Parametricism as Style—Parametricist Manifesto," London, 2008. Presented and discussed at the Dark Side Club, 11th Architecture Biennale, Venice 2008.

Chapter 11

Parametric schizophrenia
Peggy Deamer

The division in Parametrics between parametrics as form-driven scripting and parametrics as intelligence management—BIM (Building Information Modeling)—is as insistent as it is confusing. Calling the larger project of using Parametric algorithms—whether formally or informationally motivated—capital "P" Parametrics (and small "p" parametrics for the formal parametrics), this chapter offers a highly schematic analysis of the schizophrenia inherent in the Parametric enterprise.[1]

The schizophrenia is immediately evident in their *mises-en-scène*: parametric conferences are populated by young hipsters dressed in black, showing images of their digitally fabricated screens or rendered bas-reliefs; BIM conferences by older, suit-and-tie office-types explaining diagrams of complex buildings, hospital HVAC systems being a particular favorite. For those of us who experience these two dichotomous worlds, in which there is a belligerent dismissal of one for the other, diagnostic assessment feels urgent. And while the above-described schizophrenic duality relies on stereotypes, it is precisely as stereotypes—"the code of everyday signs" that help accentuate their own critiques[2]—that analysis, indeed psychoanalysis, is invited.

One of the first psychoanalysts to study schizophrenia, not as an exceptional neurosis but as an essential condition of development, was Melanie Klein. In her work on childhood psychology, Klein distinguishes between the "schizophrenic position" of the infant at the earliest stage of its development from the more mature "depressive position" that follows. In the earlier schizophrenic period, the child thinks that the good breast nurturing it is entirely distinct from the bad breast that is withheld from it. The same breast, that is, is schizophrenically divided into two part-objects—one good, the other bad. In the later depressive position, the child recognizes that the nurturing good breast it dotes on is the same as the bad breast it wants to attack. Realizing that the loved object is inherently compromised is depressing, but it is mature. The role of the analyst, in

Klein's view, is both to help the analysand, who might be stuck (as child or adult) in a "part-object schizophrenia," move to a whole-object, integrated position and also to relieve the patient from the aggression associated with part-object ecstasy and anger.

Taking the above-described dichotomous differences in the Parametric world as indications of this first, immature schizophrenic position, what follows are charts listing the contrasting attributes associated with the two sides—formal scripting and BIM—of Parametricism, contrasts subdivided according to three categories of aesthetic analysis, namely the author/maker, the object, and the audience.[3] Neither is marked as the "good" or the "bad" (breast)—this would render the analysis too undialectical; rather, each is characterized by its partiality. These charts work through the identities, projected or real, of the initial *mises-en-scène* stereotypes. In each case, in each aesthetic category, the three characterizations of the identities contrast, and it is these contrasts that require attention.

In Chart 1 (Figure 11.1), these identities can be described thus: The *parametricist* "authors" can be depicted as researchers in small, innovation-driven "labs"; maverick hipsters; and avant-gardists; the *BIM* authors as employees in large, profit-driven firms; AEC professionals; managers. In other words, in terms of how each author conceives of their role in work, the parametricists do not assume they work—they experiment, design—while the BIMers think they have precisely absorbed design AS work. The *parametricist* "objects" look something like this: forms mimicking their algorithmic systems; seamless objects making no tectonic, material or trade distinctions; and objects organized by retail logic—products; the *BIM* objects look like this: icons of functional systems; pictograms evidencing of the various construction trades; and depictions of the process logic. In other words, the objects that each identifies with are not just formally different, they occupy different dimensions of objecthood, one defined by idealism, the other by materialism. The *parametricist* "audience"—those whom the object is meant to persuade—is made up of those that are being recruited to the avant-garde—students; those post-9/11 neoliberals excited about economic prospects—Generation Y; and those organized by social media; the BIM audience, practitioners (architects); those post-babyboomers shaped by economic fear and political distrust—Generation X; and those organized by network systems. In other words, the audiences that each aims to impress and shape differ not merely politically, but in their concepts of "the public": parametrics address a general, contemporary public; BIM a "public" in the sense that anyone could be a potential "player" ready to engage in the AEC economy.

P-PARAMETRICS

	PARAMETRICS	BIM
AUTHOR	Researchers Maverick Hipsters Avant-Gardists	Employees AEC Professionals Managers
OBJECT	Algorithmic Systems Seamless Objects (unfabricated) Product Icons	Functional Systems Disjunctive Objects (fabricated) Process Icons
AUDIENCE	Students Generation Y Products of Social Media	Practitioners Generation X Products of Network Theory

Figure 11.1 Chart 1.

When fused together, these contrasting ego-defining identities point to a completeness, if we can call it that, that overcomes their partialities. Each camp offers identities that architecture is meant to embrace. Indeed, one could say, the profession needs both to mutually infiltrate each other. In other words, as analysts, we are invited to promote a "whole-object" synthesis of these two partial Parametrics. Chart 2 (Figure 11.2) indicates what this happy, "whole" position might look like: the attributes of the two kinds of "authors" make an ideal entrepreneur: savvy about positioning in the contemporary market; knowledgeable about management and business. The quintessential entrepreneur, after all, is one that recognizes branding as an imaginary *and* an economic condition. The attributes of the two kinds of "objects"—product-driven formal elegance matched with knowledge about trade and labor, staged procurement, and the tectonic difficulties these imply—would have the advantage of shifting design away from image obsession while not sacrificing aesthetics altogether. It would be informed form. The attributes of the two "audiences" would join to encompass

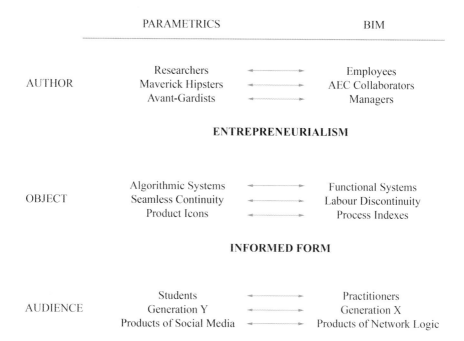

Figure 11.2 Chart 2.

the larger spectrum of what we might call the knowledge economy—socially savvy, network aligned, multigenerational. This whole-object professional identification would be very appealing in a neoliberal-meets-post-capitalist economy.

But Kleinian psychoanalysis—indeed, psychoanalysis in general—demands something more difficult than a synthesis of what are, essentially, mere symptoms; it demands an analysis of the unacknowledged desires that motivate their production. Klein and Jacques Lacan, who builds on Klein's part-object work,[4] direct us to a more critical evaluation of Parametricism's schizophrenia based on desire's underlying logic: "the desire of the other, not so much because the other holds the key to the object of desire, as because the first object of desire is to be recognized by the other."[5] And what might we say is the "Other" of Parametricism, both formal and BIM? There are many—other forms of design, other professional models (law, medicine, etc.), other forms of practice—but to go out on a limb, as analysts do, we suggest that the "Real Other" of contemporary practice is Modernism—its revolution both setting the standard for architectural relevance and avant-garde empowerment even as its failure invites repeated attempts at redemption.[6] Casting its unholy shadow on architecture by presenting an image of history that is, on the one hand, teleological and, on the other, optimizing, Modernism falsely but emphatically gives the image of the ultimate achievement of abstraction—the teleology of form—and of the ultimate triumph of efficiency—Taylorization and proficient production. We speculate here that formal parametrics embraces the first of these (formalism) while BIM embraces the second (efficiency).[7]

The consequence of this speculation is laid out in Chart 3 (Figure 11.3), again as it applies to the three aesthetic categories. Here, the descriptors are shaped by the logic of Desire, that is, by the projected image of desirability to the Other/Modernism. The *parametric* "author" is driven by the zeitgeist signification, charged with imaging the Contemporary; the *BIM* authors, valuation signification charged with imaging the metrics of architectural efficiency and profit. The parametric "object" is driven to be an *uber*-form that, while figurally flexible, is historically inevitable; the BIM object is driven to be a plethora of data, an "unhierarchical" *informe* not of detritus but of information. The parametric "audience" is driven by an identification with culture, and, as such, capitalism's superstructure; the BIM audience is driven by an identification with business, and, as such, capitalism's base.[8] In each case, the drive to kill the father figure of Modernism is aggressively pursued, even as (as we know) that figure is an empty vessel and the drive motivated by its inherent unattainability.[9]

An ideological critique of this Other-driven but still schizophrenic Parametricism is, at this point, unavoidable, moving the discussion from a Kleinian/Lacanian analysis to a Deleuze/Guattarian one. Gilles Deleuze and Félix Guattari thought that Klein was correct in her depiction of partial objects but wrong in understanding them to be a temporary "pre-Oedipal" stage *en route* to their integration in a unified ego. Instead, in Deleuze and Guattari's "schizoanalysis," "the unified ego is an illusion and objects remain throughout adulthood partial as they image the unconscious forces warring for temporary dominance in the psyche."[10] Schizoanalysis also sees the part-objects, different as they might be, as two sides of the same capitalist coin. The move then, from Chart 3 to Chart 4 (Figure 11.4), is both from a more "pure" psychoanalytic description of Parametricism's schizophrenia to one depicting the apparatus of capitalism within a psychoanalysis of Parametrics and, as well, from an unfolding dialectic "synthesis" to unembedded "ideogram," a snapshot of capitalism's underlying institutions and intelligences. As Deleuze and Guattari have indicated, the ideogram does not "mean," it shows; it does not "resolve," it reveals; it does

P-PARAMETRICS

	PARAMETRICS	BIM
AUTHOR	Zeitgeist Signifier Researchers Maverick Hipsters Avant-Gardists	Valuation Signifier Employees AEC Collaborators Managers
OBJECT	Uber-Form Algorithmic Systems Seamless Objects Product Icons	Inform-ation Functional Systems Objects of Discontinuity Process Indexes
AUDIENCE	Superstructure Students Generation Y Products of Social Media	Base Practitioners Generation X Products of Network Theory

Figure 11.3 Chart 3.

not "interpret," it lays bare.[11] Chart 4, then, depicts the ideographic status of Parametrics today. Here, the apparatus that motivates the Parametric "author," be it parametric or BIM, reveals itself to be, ultimately, the branding of proprietary design software technology—Revit, Grasshopper, Maya, etc.—and the marketing of design technology. The apparatus motivating the Parametric "object," be it parametric or BIM, is determining the look of "innovation." And, the apparatus shaping the "audience," be it parametric or BIM, is late-capitalism's sponsorship of consumption and production for their own sake.

This reductive but not inaccurate depiction of the Parametric state of affairs depicts the underbelly of Chart 2's "whole-object" optimism. What is required then is the undoing of its neoliberal aspects by awareness of and active resistance to its Chart 4 other. In Deleuze and Guattari's schizoanalysis, this awareness is linked to State deterritorialization. For them, the induction of schizophrenia in the capitalist subject is both its essential *modus operandi* and its inevitable undoing, and speculative resistance participates in the destabilizing apparatus.

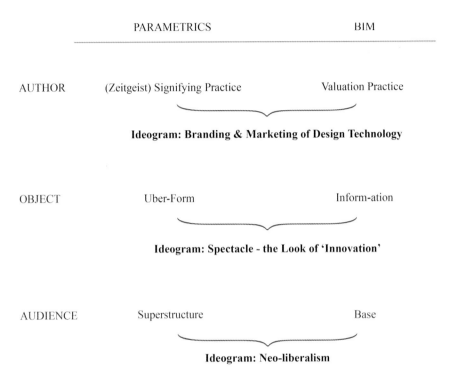

Figure 11.4 Chart 4.

The schizophrenic is inherently ready to undermine the State that has made her. For Klein and Lacan as well, the ideal object (the good breast for Klein, the *objet petit a* for Lacan) are illusions that set in motion their own undoing. In other words, any view of architectural development must not only reject a teleology of scientific, natural, or historic "progress,"[12] but foreground agency and personal resistance. If we want to have Parametricism serve a rebalancing of architecture, and architecture in turn a rebalance of production and consumption, or to align markets with social and environmental needs, architects need to deploy their Parametric power strategically.

Chart 5 (Figure 11.5) depicts the ground that is opened by a whole-object fantasy (Chart 2) and its territorializing pull (Chart 4) to imagine a deterritorialized postcapitalist outcome. In this, the architectural "author," oscillating between branding and entrepreneurialism, can, if nothing else, demand that the reality of the proprietary software companies conform to the claims of democratic, affordable, and unmonitored information sharing. The architectural "object"

P-PARAMETRICS ←——→ ARCHITECTURE

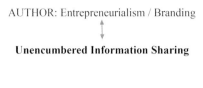

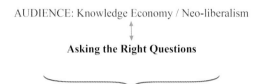

Figure 11.5 Chart 5.

would announce that its spectacle is not merely visual but spatial, that its intelligence was not merely economic but aesthetic, where "aesthetic" refers to the profound way in which spaces, materials, proportions, structure, stability, and light convey architecture's message. The "audience" constructed by the knowledge economy would demand information about who really needs and profits from Parametric work and the objects it yields. As Peter Drucker, the 1950s management guru, indicated, the important issue for us architects is not to find the right answers but to ask the right questions.[13]

The deterritorialized architectural profession could, indeed, enter the knowledge economy as a full but revolutionary player. This would entail a number of things: being valued for our expertise in building procurement and building performance; structuring a new basis for calculating fees based not on percentage of construction (piece-work) or hourly fees (time-work), but on the value we create and the long-term money we save our clients; proving our expertise as spatial organizers and public producers; moving away from developer-driven, free-market systems toward a more democratic ordering of the city; working toward the establishment of a universal wage across the discipline and, eventually, society.[14]

The schizophrenia currently experienced by the architectural profession in general and by Parametricism in particular—one that is unhealthy, un-self-reflexive, and incomplete in both characterizations—will not be "healed"; rather, as this initial analysis promotes, it morphs into something more aggressively sly and consciously strategic.

In this, Parametricism's role in architecture's "development" is neither necessary nor sufficient; architecture's entry into capitalism's knowledge economy could occur without this semiotic system. But, in facilitating an epistemological framework that demands informational connectivity, allows disciplinary hybridization, encourages authorial collaboration, and structures fiduciary responsibility, it is a powerful ingredient in undoing existing dysfunctionalities[15]. This framework is not only one that organizes a better world but restructures our own discipline. Indeed, we need to recognize that there cannot be a division between what we produce "for the world" and how we produce it institutionally. As long as architecture as an institution has and supports horrific labor practices, does not share its meager rewards fairly, and is led largely by white males, it will never produce the society that does not also have these limitations. For me, the inspiration of Parametricism, schizophrenic or otherwise, is its radical reorganization of practice itself. Deterritorialized workers, unite!

Notes

1. Formal parametricism is associated with the work of, among many others, Greg Lynn, Hernan Diaz Alonso, & Zaha Hadid/Patrik Schumacher. Its output is most consistently published in *AD Design Magazine* (London: Wiley-Academy) from 2002 to the present. Information-based parametricism is not associated with particular designers, which is consistent with this side of the schizophrenia; that is, it's being associated with production, not design firms; SHoP (also associated with formal parametricism) and Kieren Timberlake are well-known firms that are the exception. The publications associated with knowledge parametrics are Chuck Eastman, Paul Teicholz, Rafael Sacks, & Kathleen Liston's *BIM Handbook: A Guide to Building Information Modeling for Owners, Managers, Designers, Engineers, and Contractors* (Hoboken, NJ: Wiley and Sons, 2011, second edition) and eds. Douglas Noble & Karen Kensek's, *Building Information Modeling: BIM in Current and Future Practice* (Hoboken, NJ: Wiley and Sons, 2014). Likewise, the RIBA has aggressively pushed the use of BIM in Britain, publishing four versions of their National Building Specifications (NBS) on BIM's effect and potential in the AEC industry.

2. Pierre Klossowski, in his work on Nietzsche, Deleuze, and Guattari, developed his ideas on the stereotype in his "On the use of stereotypes and the censure exercised by classical syntax," in *Protase et Apodose*. As Daniel W. Smith, in his "Translator's Preface" to Klossowski's *Nietzsche and the Vicious Circle* (Chicago: University of Chicago Press, 1997) writes, "Klossowski (…) speaks of a 'science of stereotypes' in which the stereotype, by being 'accentuated' to the point of excess, can itself bring about a critique of its own gregarious interpretation of the phantasm: 'Practised advisedly,' the institutional stereotypes (of syntax) provoke the presence of what they circumscribe; their circumlocutions conceal the incongruity of the phantasm but at the same time trace the outline of its opaque physiognomy." Found at http://www.google.com/url?sa=t&rct=j&q=&esrc=s&source=web&cd=2&ved=0CCIQFjAB&url=http%3A%2F%2Fwww.sauer-thompson.com%2Fessays%2FSmith%27s%2520Preface.doc&ei=oICrU-f9OMqryAS0soLgBw&usg=AFQjCNE4SVw0nIaKCkhdJJnTruFLecXXDQ&sig2=VmeG2HAwKEITY7fHvkWnqQ&bvm=bv.69837884,d.aWw [accessed May 28, 2014].
3. This division is less "known" than it is deduced from psychoanalytic writings on aesthetics. The common use of psychoanalysis in literature or painting is to analyze the author's psyche (van Gogh) or the characters depicted (Hamlet). Less common is an analysis of the aesthetic object. Adrian Stokes, a follower of Melanie Klein, is one of the few to make this his life's work.
4. Lacan was influenced by Klein and supported her in her battle against Anna Freud for dominance in childhood psychoanalysis. He agreed with Klein that attachments were, in distinction to Freud's view, to external objects, not to one's sexual organs. Her description of the part-object position influenced Lacan's own of "the Imaginary." But, ultimately, he moved past Klein to reassert his more essential attachment to Freudian analysis.
5. Jacques Lacan, *Écrits: A Selection* (London: Tavistock, 1977 [1959]), p. 58.
6. A dissertation could be written on this claim, but, 100 years after its birth, it is strange that we still define ourselves, with post-post-modernity, by its terms. Indeed, it seems to me that this is singularly true in architecture, where Le Corbusier and Mies van der Rohe are still the iconic figures that no figure in modern painting, not even Picasso, or no figure in modern literature, not even Proust, can match.
7. These two characterizations of Modernism—as wholly abstract and as wholly efficient—are themselves myths that both Modernism and post-Modernism's critique of Modernism have indulged in.
8. I realize that these two terms—"superstructure" and "base"—are highly disparaged in current neo-Marxism. But they capture a difference that is poignant here: that the one side addresses culture (superstructure) while the other addresses our pocket-book; the economy (base).
9. The drive and desire are both defined, although differently, by the unattainability of their object. When we obtain the object that we desire, we do not achieve satisfaction, only another object of desire. Through the drive, the subject finds satisfaction in the repetition of the failure and loss that initially constitutes this.

10. Eds. Daniel W. Smith & Henry Somers-Hall, *The Cambridge Companion to Deleuze* (Cambridge: Cambridge University Press, 2012), p. 329.
11. One distinction between Deleuze/Guattari and Lacan is that the former do not privilege the linguistic system of the unconscious and instead, in a manner close to Klein's depiction of a psyche dominate by part-object images (although hers were not culturally determined), they assume a semiotic system made of pictograms that reveal their ideological origins.
12. The most intellectually distressing part of Patrik Schumacher's defense of (formal) parametricism is the Hegelian teleology that it subscribes to—the idea that there is continuous progress in architectural form-making that leads inevitably to parametric design.
13. See Peter F. Drucker's *The Effective Executive: The Definitive Guide to Getting the Right Things Done* (New York: HarperCollins Publishers, 1967). See also his later work, *Post-Capitalist Society* (New York: HarperCollins Publishers, 1993).
14. Manuel Shvartzberg articulated these demands in the framing of the Architecture Lobby manifesto, and I am indebted to his contributions to that work. That manifesto and his reading of it at the 2014 Venice Biennale can be viewed at http://architecture-lobby.org [accessed October 30, 2014].
15. I am indebted to Paolo Tombesi for much of these potential positive characterizations. See his "On the Cultural Separation of Design Labor," *Building (in) The Future: Recasting Labor in Architecture*, (New York: Princeton Architectural Press, 2010), p. 117–136.

Chapter 12

The architecture of neoliberalism
Teddy Cruz

There continues to be an inability to envisage the problems facing our societies today in a political way. Political questions always involve decisions, which require us to make a choice between conflicting alternatives. This incapacity to think politically is due to the uncontested hegemony of liberalism, which has re-installed a rational and individualistic belief in the availability of a universal consensus as the basis for liberal democracy, negating antagonism and conflict. This kind of liberalism is unable to adequately grasp the pluralistic nature of the social world, with the conflicts that pluralism entails: conflicts for which no rational solution can ever exists.

The belief in the possibility of a universal rational consensus has put democratic thinking on the wrong track. Instead of designing the institutions, which through impartial procedures would reconcile all conflicting interests, the task for democratic theorists and politicians is to envisage the creation of a vibrant

---

I would like to thank Matthew Poole and Manuel Shvartzberg for inviting me to participate in this project and for their helpful editorial suggestions; and Fonna Forman for our collaboration on many of the ideas that emerged in this chapter.

"agonistic" public sphere of contestation where different hegemonic political projects can be confronted.
—Chantal Mouffe, *The Return of the Political*[1]

Introduction: Parametricism and the (neutral) politics of consensus

One central provocation that motivates my reflections here is the declaration by Patrik Schumacher during the conference The Politics of Parametricism, when he indicated that the goal of parametric architecture is to give aesthetic order to the visual messiness of the neoliberal city. The suggestion that the neoliberal city is simply an object that needs to be stylistically unified ignores the fact that the actual havoc that neoliberalism has exerted upon the contemporary city in the last decades of economic boom is not just a visual phenomenon but an actual institutional process that produced a violent blow to our economic, social and natural resources.

While my first reaction, then, is to say that there is nothing political about parametricism, it would be naive not to acknowledge the crypto-political dimension of this project as it smoothly aligns itself with the power of a neoliberal political economy of urban growth that has been characterized by an antipublic agenda, engendering unprecedented urban asymmetry and socioeconomic inequality today. This is probably the main complaint I have about this seemingly neo-avant-garde architectural agenda today: its overt indifference to the societal and urban conflicts that could in fact be the critical content to investigate and deploy new architectural parameters that can problematize the politics of aesthetics and close the gap between social responsibility and artistic experimentation—a fundamental tenet of the historic avant-garde.

Not that architecture design should give a damn about resolving the crisis of urbanization today—I am aware of the limitations of any specialized field; but in the last hundred years, the most salient avant-garde movements in architecture have always engaged pressing societal issues and their formal and aesthetic consequences (from Le Corbusier's foundational CIAM battle-diagram that rallied its members to engage in social housing issues, to Constant's search for an architecture organized around relational, social contingencies).

But, this indifference to the socioeconomic and political material that could in fact "complicate" architectural form today has also been the DNA of *autonomy* in architecture, whose recurring avant-garde utopian dream throughout history has always been to give formal order to the chaos of social difference by imposing structural and compositional strategies that somehow will bring political, cultural and aesthetic unity to a society run amok. While in some ways I agree with recent political stances in architecture to return to a notion of autonomy in order to resist the aesthetic relativism behind the speculative commercial logic of hyper-capitalism, I am critical of the nostalgic return to a top-down autonomous and self-referential language as the only way out of this continued "postmodern nightmare." And, while I also agree with how critiques of autonomy have been oriented toward the bottom-up consumerist politics of capitalism, I equally condemn its abandonment of bottom-up social movements and the contested space between the public and the private, whose antagonism is at the center of the political in urbanization today. After all, in the absence of a progressive welfare state that can support the reinstating of an architectural public and social agenda at a massive scale, who is going to build these top-down architectural dreams today if not antidemocratic governments, autocratic dictatorships and the corporate neoliberal economic power of privatization?

For this reason, I consider Parametricism the official architectural face-lift of neoliberalism, as it unifies and materializes the universalist consensus politics of neoliberal global capital, into an apolitical formalist project of beautification, whose relentless homogeneity hides any vestige of difference, and the conflicts that are at the basis of today's urban crisis. Parametricism, just like neoliberalism, negates the political in its antagonistic dimension by ignoring the multiplicity of socioeconomic relations that should inspire new architectural paradigms today.

The complexity of our contemporary metropolitan condition should in fact solicit a more experimental architecture, which can only emerge from engaging with and negotiating today's urban crisis as the radical context from which to produce new aesthetic categories that problematize the relationship between the social, the political and the formal.

Neoliberal violence: Urban crisis and the death of the public

It is obvious by now, then, that the celebrated metropolitan explosion of the recent economic boom also produced, in tandem, a dramatic project of marginalization. This has resulted in the unprecedented growth of slums surrounding major urban centers, increasing the urban conflicts of uneven urbanization. This urban asymmetry, which is at the center of today's socioeconomic crises, also brought with it the incremental erosion of a public imagination, as many governments around the world enabled the encroachment of the private into the public.

This personal critique on parametricism is primarily a critique of its indifference to matters of concern in the contemporary city and its withdrawal from the politics of the public, at a moment when the *public institution* is collapsing as an ideal within a political climate still driven by inequality, lack of institutional accountability and economic austerity. In other words, as the longevity of the *top-down public paradigm* is in question today, we need urgently to search for alternatives and seek a more functional manifestation of public thinking and action at other scales and within community-based dynamics: a bottom-up public? Plus, we urgently need to search for alternative roles that architecture can play in supporting such new paradigms. The questions must be different questions if we want different answers. This is why one of the most relevant and critical challenges in our time is the problem of how we are to restore the ethical imperative among individuals, collectives and institutions to coproduce the city, as well as new models of cohabitation and coexistence in anticipation of socioeconomic inclusion.

Rethinking the public cannot begin without exposing the controversies and conflicts that define the present moment's unprecedented socioeconomic inequality. In fact, decoupling the public from the imperative of socioeconomic equity only risks romanticizing our notions of the public, perpetuating the de-politicization of this urgent topic from our artistic fields and their practices. As a point of departure, this is a political project that contemporary architecture must engage. Today, as urban designers, we cannot begin any conversation about the future of the megacity without critically understanding the conditions that have produced the present crisis.

Since the early 1980s, with the ascendance of neoliberal economic policies based on the deregulation and privatization of public resources, we have witnessed how an unchecked culture of individual and corporate greed has yielded unprecedented income inequality and social disparity. This new period marked by a lack of institutional accountability and illegality has been framed politically by the erroneous idea that democracy is *the right to be left alone*, a private dream devoid of social responsibility. But the mythology of this version of free-market trickle-down economics, assuring the public that if we forgive the wealthy their taxes all of us will benefit and one day become as rich, has been proven wrong by the undeniable evidence that political economists Saez and Piketty[2] have brought to light. They have exposed that during both the Great Depression of 1929 and today's economic downturn we find both the *largest* socioeconomic inequality and the *lowest* marginal taxation of the wealthy. These are instances when the shift of resources from the many to the very few has exerted the greatest violence to our public institutions and our social economy. The polarization of wealth and poverty in the last decades has been a direct result of the polarization of public and private resources, and this has had dramatic implications in the construction of the contemporary city and the uneven growth that has radically increased territories of poverty.

This hijacking of the public by the private in fact has been mobilized by a powerful elite of individual and corporate wealth, who in the name of free-market economic policies has enjoyed the endorsement of federal and municipal governments to deregulate and privatize public resources and spaces of the city. This has prompted many planning and economic development offices to "unplug" from communities and neighborhoods at the margins of the predictable zones of economic investment, resulting in the uneven urban development that has characterized many cities in the world, from Shanghai to New York City. This retreat of the institutions of governance from public investment has resulted also in the erosion of public participation in the urban political process as many communities affected by this public withdrawal have not been meaningfully involved in the planning processes behind these urban transformations, nor benefited from the municipal and private profits they engendered.

An argument can be made, however, that broad structural political and social changes are possible. Such changes have occurred in certain moments in history, when the instruments of urban development were primarily driven by an investment in the public—and supported by the institutions of architecture and design; for example, the emergence of the New Deal in the United States after the 1929 economic crisis, or the postwar Social Democratic urban politics in Europe. But today's crisis and its conditions are exponentially more complex, as the consolidation of exclusionary power is not only economic but also political, driven by one of the largest corporate lobbying machines in history, the elements of which have subordinated collective responsibility to serve individual interests, dramatically changing the terms by polarizing institutions and publics, wealth and power, and misallocating natural, social and financial assets in unprecedented ways.

I am not suggesting here that an architecture agenda such as parametricism should be expected to take responsibility for these crises. In fact, it is often the response from many architects that are part of this new paradigm that the critical issues affecting us today—ranging from global warming to inequality—are not *their problem*, because architecture is not social-work, nor should it have a political position. I am not endorsing here the clichés of the generic project of social justice and activism in architecture, which usually translates into symbolic problem-solving relief efforts that unfortunately do nothing to interrupt the backward policies that produced the crisis in the first place, and end up only antagonizing design and aesthetics. Nor I am advocating for "political architectures," in fact, but for *the construction of the political itself*, and I believe that our creative fields can engage this challenge, as the problems of urbanization cannot be tackled with buildings only, but through fundamental institutional shifts; a new socially responsible political ground that can yield more socioeconomically complex and pluralistic environments.

Smooth skins and the camouflaging of disaster

Much of my research on the trans-border urbanisms that have informed my practice first began as a desire to critically observe and trace the specificity of such conflicts inscribed in the territory of the San Diego/Tijuana border, one of the most contested borders in the world. Not only has observation of this region enabled a practice of research, a sort of projective forensics of the territory, but it also revealed the need to expand narrow notions of architectural design. The revelation was simple: without retroactively tracing and projectively altering the backward exclusionary policies that have been constructing many geographies of conflict worldwide in the last few years, whether local urban zones of conflict or larger border regions, our work as architects will continue to be a mere camouflaging (and decoration at best) of selfish, oil-hungry urbanization from China to Dubai, to New York, to San Diego.

It has been disheartening, for example, to witness how the world's architecture intelligentsia—supported by the strong economy of the last years—flocked *en-masse* to the United Arab Emirates and China to help build the dream-castles that would catapult these enclaves of wealth—and even fascist autocracy, in some instances—as global epicenters of urban development. But, other than a few architectural interventions by high-profile protagonists whose images of sameness have been disseminated widely, no major ideas were advanced here to resolve the major problems of urbanization today, which are grounded in the inability of institutions of urban development to more meaningfully engage urban informality, socioeconomic inequity, environmental degradation, lack of affordable housing, inclusive public infrastructure and civic participation.

In the context of these shifts, we are paralyzed across sectors, silently witnessing the consolidation of the most blatant politics of unaccountability, the shrinkage of social and public institutions and not one single proposal or action that can suggest a different approach with different arrangements. At the moment when such collective creativity produced by some of the most important architecture manifestos in the last decades could have deployed their intelligence in redefining the contemporary city, charting a new era of urbanization, many of these proposals simply became mere caricatures of change, preoccupied only with the *look* of the city, expressed by a collection of discreet architectural icons, whose smooth skins made them into commodities of beautification, subordinated to the most greedy version of a free-market economic model.

In this case, parametric architecture in its pursuit to homogenize the city with one stylistic gesture is not that different from the metropolitan homogenization produced decades ago by the worst version of an *international style*—an agenda that by the late 1950s had already emptied itself of the postwar modernist political aspirations to reconnect architecture and the social, turning instead into the privileged formal resolution of an international and homogeneous urban development recipe defined by a "minimum investment maximum profit" paradigm: the privileged style of corporate power. So, before being an economic and environmental crisis, ours is primarily a cultural crisis resulting in the inability of the institutions or urbanization to question their ways of thinking, their exclusionary policies, the rigidity of their own protocols and silos, and in the case of architecture, its incapacity to reimagine itself as a social and political medium to produce counter narratives to these totalizing trends, i.e., other forms of economic development.

In fact, it is not enough for architecture and urban design to camouflage, with hyper-aesthetics and forms of beautification, the exclusionary politics and economics of urban development; at this moment it is not buildings but the fundamental reorganization of socioeconomic relations that must ground the expansion of democratization and urbanization. It is necessary, then, that the political specificity shaping the institutional mechanisms that have endorsed today's uneven urban development must be the catalyst for new architectural paradigms. In other words, the critical knowledge of the conditions themselves that produced the global crisis should be the material for architects in our time, making urban conflict the most important creative tool to reimagine the city today, and the generative platform from which to develop policy proposals and counter-urban development strategies, helping us to reimagine how the surplus value of urbanization can be redirected to the social.

Ultimately, it is irrelevant whether urban development is wrapped in the latest parametric, morphogenetic skin, neoclassical "new-urbanist" prop or environmentally sensitive photovoltaic panels, if all such approaches continue to camouflage the most pressing problems of urbanization today. Without altering the exclusionary political and economic power that has produced the current crises, the future of the city will continue to be subordinated to the visionless environments defined by the bottom-line urbanism of the developers' spreadsheets and the political economies of an uneven urban growth sponsored by neoliberalism everywhere.

Just like the future ecological disaster that is currently camouflaged by the picturesque pastoral identity sponsored by the oil-hungry suburban sprawl in many edge-cities in the world, the smooth skin of parametric architecture, in its relentless homogeneity and antisocial agenda, might be hiding an urban-disaster-in-the-making. It is as if history does not exist—we all know how certain architectural utopian dreams can soon transform into dystopian nightmares. Can these masks of vanity be the innocent political symbols of an evolving mono-cultural and antidemocratic urban ethos, where all conflicts come to an end, and a class-less society lives happily ever after?

The informal algorithm and the new political economy of form

If parametric architecture did to the political what it does to form, maybe then we could open up new spaces of operation as algorithm-based morphogenetic-inspired architecture could deploy its intelligent systems thinking in the production and representation of new political and economic "envelopes" within which bottom-up spatial emergencies and top-down organizing logics could collaborate.

In fact, recent shifts in the last two decades, from linear thinking to systems thinking in science and many other related fields, show a move from the type of silo-thinking that fields of specialization have perpetuated to a more complex ecology of inter-dependences, embedding scientific and formal research into a more complex cultural context. The hope is that Parametricism in architecture could complicate itself by more comprehensively engaging the sort of systems thinking that is (seemingly) at the base of its generative formal processes. The problems of the contemporary city are systemic problems that need a deeper ecological thinking. Fritjof Capra's notion of "deep ecology" illustrates this very well when he suggests the need to challenge traditional understandings of ecology: conventionally, the ecology of a bicycle pertains to the functional relationship of the parts of the bicycle to themselves. A deeper ecology of this bicycle would have to also ask who produces this bicycle? Where? And what is its cultural application? As the cultural consciousness of bike riding varies in California, India and Denmark.

But, instead, the algorithms that inform parametric architecture's formal compositions suffer from the "precariousness of omission," as the selective set of parameters that are chosen for the assembly of self-organizing patterns, volumes and effects (which can, in any given context, be the same, from a door-knob to an entire city!) simply respond to an amazingly reductive combinatory set of formal attributes that always exclude the complexity of other vectors of force that are exerted in the metropolitan battlefield. Anticipating the triviality of these autonomous formal games in years to come, Rem Koolhaas wrote in the late 1980s about how the former Pan Am building, above Grand Central Station, despite its "clumsiness" and bad details, was the most revolutionary building in New York City as it had intervened into the contested space of air-rights, enabling itself to float over the historic landmark by exploiting the politics of verticality, not through decorative, exterior form, but in its sectional complexity and programmatic flows.

In essence, parametricism's focus on material systems in such self-referential ways today (without any relationship to the political economies these materials can infiltrate and shape, without any awareness of the modes of production and labor that go into their execution, without any attention to the deeper ecological context of the metropolitan and the rural today, and without any acknowledgment of the social contingencies that always usurp the autonomy of exclusionary form) amounts to yet another of those a-historical moments when the pendulum of architectural ideology shamelessly tilts to the narcissism of an aesthetics-for-aesthetics-sake agenda at the expense of a mutating world around it.

Instead, the new political economy of form today is found not in architecture's self-referentiality but in the construction of new socioeconomic and political "envelopes" from which new formal configurations can emerge. This also suggests the need to produce new clients, and new sites of intervention and research. One of the most critical sites of intervention, for example, is the developer's *pro-forma*, the economic spreadsheet that gives "form" to a building. How to intervene in its organizational logics of financing and resources? Or, how to intervene in the gap between the formal and the social: How to simultaneously design space and the protocols that can assure inclusion and a more pluralistic democracy? In essence: How to be more comprehensive and acknowledge the parameters that have been absent from those self-organizing formal logics, namely the political, social and economic domains that have remained peripheral to design today? Here, the social and economic contingencies found in informal urbanization can give us clues about some of the problematics missing from parametricist formalism.

Conclusion: The informal algorithm manifesto

The informal algorithm is the unit from which to generate other ways of constructing the city by mediating between top-down and bottom-up dynamics: in one direction, how specific, bottom-up urban alterations by informal contingencies can have enough resolution and political agency to trickle upward to transform top-down institutional structures; and, in the other direction, how top-down resources can reach sites of marginalization, transforming normative ideas of infrastructure, by absorbing the creative intelligence embedded in informal dynamics. This critical interface between top-down and bottom-up resources and knowledges is essential in the restoration of the linkages between government, social networks and cultural institutions to reorient the surplus value of urbanization toward a public imagination, from which a new infrastructural way of thinking can frame the future of the city:

1. To challenge the autonomy of buildings, often conceived as self-referential systems, benefiting the one-dimensionality of the object and indifferent to socioeconomic temporalities embedded in the city. How to engage instead the complex temporalization of space found in informal urbanization's management of time, people, spaces and resources?
2. To question exclusionary recipes of zoning, understanding it not as the punitive tool that prevents socialization, but instead as a generative tool that organizes and anticipates local social and economic activity at the scale of neighborhoods.
3. To politicize density, no longer measured as an abstract amount of objects per acre but as an amount of socioeconomic exchanges per acre.
4. To retrofit the large with the small. The micro-socioeconomic contingencies of the informal will transform the homogeneous largeness of official urbanization into more sustainable, plural and complex environments.
5. To reimagine exclusionary logics that shape jurisdiction. Conventional government protocols give primacy to the abstraction of administrative boundaries over the social and environmental boundaries that informality negotiates as devices to construct community.
6. To produce new forms of local governance, along with the social protection systems that can enable guarantees for marginalized communities to be in control of their own modes of production.
7. To enable more inclusive and meaningful systems of political representation and civic engagement at the scale of neighborhoods, tactically recalibrating individual and collective interests.
8. To rethink existing models of property by redefining affordability and the value of social participation, enhancing the role of communities in coproducing housing, and enabling a more inclusive idea of ownership.
9. To elevate the incremental low-cost layering of urban development found in informal urbanization in order to generate new paradigms of public

infrastructure, beyond the dominance of private development alone and its exorbitant budgets.

10. To mobilize social networks into new spatial and economic infrastructures that benefit local communities in the long term, beyond the short-term problem solving of private developers or institutions of charity.
11. To sponsor mediating agencies that can curate the interface between top-down, government-led infrastructural support and the creative bottom-up intelligence and sweat-equity of communities and activists.
12. To close the gap between the abstraction of large-scale planning logics and the specificity of everyday practices.
13. To challenge the idea of public space as an ambiguous and neutral place of beautification. We must move the discussion from the neutrality of the institutional public to the specificity of urban rights.
14. To layer public space with protocols, designing not only physical systems but also the collaborative socioeconomic and cultural programming and management to assure accessibility and sustainability in the long term.
15. To enable communities to manage their own resources and share the profits of urbanization to prevent gentrification.

Notes

1. Chantal Mouffe, *The Return of the Political* (London: Verso, 2006); © Chantal Mouffe and Verso, London, 2006, reproduced with kind permission of Verso, 2014.
2. Emmanuel Saez and Thomas Piketty, "Top Incomes and the Great Recession: Recent Evolutions and Policy Implications," *International Monetary Fund Economic Review* # 61 Vol. 3, (2013), pp. 456–78. Found at http://eml.berkeley.edu/~saez/piketty-saezIM-F13topincomes.pdf [accessed October 13, 2014].

Chapter 13

Parameter value
Phillip G. Bernstein

1

In the fall of 2013, San Francisco witnessed one of the most surprising and unlikely come-from-behind victories in the history of sports. Team Oracle USA overcame an eight-race deficit to beat New Zealand 9–8 in the Americas Cup sailed in the icy waters of San Francisco Bay. Oracle CEO Larry Ellison, known in Silicon Valley as a brilliant if ruthless tactician, was not actually on the boat that won, but was both its sponsor and mastermind. He had assembled experienced crews that built and manned the super-high-tech catamaran—which often reached fifty knots barely touching the water as it skimmed across the waves—plus, an additional, unseen, crewman who likely made the difference between ignominy and victory: big data.

All America's Cup boats are produced at the leading edge of technology with the most advanced manufacturing methods, digital modeling, extensive pre-race simulation, and calibration. The built artifact has its analytical doppelganger back in the design lab in the form of various parametric simulations, likely terabytes of data in various formats, an approach that a database company such as Oracle no doubt finds appealing. But Oracle's strategy took big data out on the water itself. The physical boat—festooned with 300 sensors—and its analytical digital counterpart were in constant conversation during the races, with huge amounts of telemetry returning to shore and allowing the team, both on and off the water, to measure, recalibrate, and tune the boat's behavior in real time, adjusting the values of various performance parameters to optimize the boat's performance. A constant feedback loop between simulation and reality gave the crew minute instructions on how to optimize their machine toward one end: going as fast as possible. You might say that Oracle's team rode a wave of parameters to victory—an ending totally appropriate to the aggressive Ellison and the company he leads. No expense was spared in this effort, with Oracle's challenge costing him over $100 million.

Three decades ago design and construction technology was more allied to methods and approaches that would be familiar to Columbus sailing to the new world with sketchy maps and astrolabes. Before the digital era in our industry, largely considered to have begun in earnest when personal computers and drafting software became available in the mid-1980s, experience, rules of thumb, and abstraction were the primary tools of design delineation, communication, and execution. Designers and builders were more closely allied, buildings less complicated, legal considerations largely distant, and efficiency of depiction in the form of traditional orthographic drawings like plan, section, and elevation sufficient for the task at hand. Architectural drawings were a strategy for depicting and abstracting a very large thing (a building) via very small things (sheets of paper) using techniques like planometric projection, scale, and graphical metadata like dimensions and textual annotation. Only rarely would a given building component be represented at full scale and materiality in these drawings, as scale and abstraction were the fundamental principles of architectural drawing necessary to explain the desired result.

Historian Mario Carpo argues that mechanical representation of building design by abstract diagrams on sheets of paper was essentially a problem of limited available computational power.[1] Buildings are very big and complex entities comprising an enormous number of pieces, parts, relationships, and connections. Drawings were a very efficient way of "storing and managing" all that data by abstracting them into a particular representational language. But as "stored" on the paper, the resulting information was impossible to query or manipulate in any robust or meaningful way beyond simple counting and measuring. Drawings were the proxies for the conceptual models that Alberti asserted were separate in the mind of the architect from the material result by the builder.[2]

The drafting instruments that created these depictions were hardly different from their seventeenth-century counterparts, and the underlying mathematics—arithmetic and geometric "analytical models" of the nondigital age—were implicit, implied by the length, organization, and annotation of building abstractions that were memorialized in drawings. Engineers responsible for more performative aspects of building (like structure and building systems) relied on printed tables of standard data, hand-managed equations, slide rules, and large safety factors that accommodated the inherent imprecision of the resulting predictions of behavior. Architects largely delegated all such questions to them accordingly, maintaining for themselves control of the aesthetic and compositional, while delegating obligation for technical performance to their consultants. But soon computers would, in part, come to the rescue.

A first wave of digital integration began with the translation of some techniques from hand methods to their computerized counterparts. Slide rules give way to calculators, drafting tables to copies of AutoCAD, and steel manuals to early structural computation programs. While the results of these processes were the same—even more drawings produced with digital accuracy—two important representational shifts that presaged so-called parametricism emerge. First, even though methodological conceits like "paper space" and "layers" enabled efficient drawing production, the underlying representation of the building in the form of electronic lines, arcs, and circles was made at full scale and full precision, creating the earliest version of a geometric "model" (in some sense) of the building. That model was created by direct manipulation of mathematical vectors, and the resulting electronic drawing was then scaled and projected by plot onto paper. Second, and more importantly, the digital vectors that comprised the scale drawings of the building also created an explicit mathematical and virtual definition of the building, an analyzable model of sorts that could be manipulated by both mouse and equation.

The manipulation of the lines, arcs, and circle of CAD evolved, over time, to the concept of blocks: integrated, connected bunches of geometry that could be repeated, reflected, or scaled and that made the management of large swaths of digital geometric data more palatable and (allegedly) drafting more efficient. As computational power and storage capacity of desktop computers increased via Moore's Law,[3] blocks could give way to more robust chunks of digital geometry called objects that carried with them certain parametric and other data allowing them to be somewhat self-aware of their role in a drawing and therefore contextualized by those parameters to fit a certain set of circumstances. For example, the graphic collection of information representing a "door" would include a scalable plan drawing of the door and its swing, insert itself into the similar representation of its wall, and report its characteristics to the chart that was the door schedule; it could be counted and coordinated more rapidly into the overall drawing set. But because much of the information in that drawing set—things like floor patterns, details, fixtures and furniture, or ceiling systems—was still entirely graphic, the parametric intelligence of object-oriented CAD was limited at best and largely deployed in the service of better technical drafting. It was easy to copy, block, and scale these smart drafting elements. Ironically, the age of CAD brought with it an explosion of additional drawing in architecture, particularly in production drawings, and why not? The incremental cost of digital drafting space was essentially zero, and output mechanized via plotters.

Eventually two-dimensional drafting platforms like AutoCAD would evolve to allow three-dimensional representation, but still aimed at the ends of drafting.

In the parallel universes of manufacturing and animation, however, three-dimensional geometric modeling had accelerated into the mainstream, and with it tools like Rhino, Pro Engineer, and Maya, each of which brought with them the ability to create geometry and access it by both direct manipulation and by formula; the equation that generated the geometry was bi-directionally accessible to a clever programmer who could "extract" the length of a line or the radius of an arc, manipulate it by script, and regenerate the resulting geometry. Parametric form generation, or the use of scripted mathematics that accessed the now completely available "analytical model" of the geometry, was born, and with it a generation of what Dean Robert Stern of Yale University has endearingly called "The Blobmeisters." Scripted manipulation of geometry parameters blossoms to this day when designers create algorithms that access the analytical geometric model (the now explicit math) and generate forms based on those routines. As Dean Stern suggested with his gentle rib, a great deal of that work was inspired by the non-rectilinear, sinuous forms and shapes that would result. And, since building component providers could access that same geometry to power computer-controlled fabrication, the blobs that appeared on screen soon appeared on the construction site. When it is easy to draw a curve, how much harder can it be to build it? Apparently, not much. Even in mainstream design and construction, AutoCAD made drafting rounded shapes so easy that it influenced the expressive language of institutional architecture of the 1990s; drawing those curves and cutting the resulting extruded shapes were much less of a challenge.

Object-oriented CAD and parametric geometry engines presaged the advent of a disruptive technology in the building industry, namely building information modeling (BIM). BIM took its cues from the technology of manufacturing design that had for years created detailed "solid models" of airplanes and cars that including geometry, materiality, behavior and parametric characteristics. The polarity of data in such models was reversed, in the sense that drawings were derived from the solid models themselves rather than being authored directly by the software.

Architects, engineers, and builders had inherited much of their digital toolbox from other disciplines, and manufacturing modelers in particular were in the main unable to handle the large scale, complexity, and degrees of abstraction necessary to represent a building, and particularly the arcane language of architectural drawing. Thus, various vendors created a generation of purpose-built tools known as "building information modelers." BIM platforms, which began to mature in the mid-2000s, provided the first real full "simulative" representations of buildings based on data structures and an underlying digital epistemology of architecture. Drawings, renderings, schedules, and projections could be generated as reports from a database that included geometry, annotations, material and behavioral data, and parameters to control various aspects of the represented element. As important, the elements exist in parametric relationship to one another (allowing systems of elements to be manipulated as relational groups). The data can be accessed for such manipulation by increasingly mature software "ports" called application programming interfaces. With the advent of BIM, the building industry obtained its first tools for building behavioral simulation, complete with parameters and access, representing both the intent and the result of design acts and the ability to manipulate those elements by algorithmic routine in ways more robust and provocative than merely bending flat planes into curves.

In broad strokes, it is possible to see the evolution of digital technology in building representation in three successive frames of reference. In the first "era," manual drafting techniques were supplanted by electronic drawings, but drawings were still the primary means of creating the representation of the design itself, and documentation remained the main focus of the tools. It has been argued, in the first installment of a recent exhibition tracing the history of computation in architecture,[4] that parametricism as a design strategy was "invented" by Peter Eisenman in concert with Chris Yessios in their work on the competition entry for the Biozentrum at Goethe University. In that project, primitive scripted computation was combined with computer graphics as scheme generator, creating what curator Greg Lynn characterized as the first parametric project at the dawn of computerized drafting.

A second disruption came with BIM, where drawings were no longer the object of the representational instruments, but rather an artifact extracted when necessary from the parametric model of the building. That parametric model becomes a prototype for the finished building and can be visualized and measured in detail. The emergence of algorithmic techniques, seen combined with both BIM and the computational power of the cloud, brings us to the beginning of a third era of digital contextualization, where various systems of relationships, physical components, large data structures, and other digitized information connect and combine to virtualize all of the systems of building. The implications of the era of context will depend as much on the power and efficacy of digital tools as the designer's will to change the nature of the design process itself in their use. Operating in the context of a wide array of parameters made accessible by integrated digital representation, architects can refactor, reprioritize, and restructure both the fundamental heuristics and the results of the design process itself.

2

In his 1970s treatise *The Architecture Machine*, architecturally trained computer scientist Nicolas Negroponte posited that digital technologies first mimic the processes that they are designed to replace, then extend them, and eventually disrupt them completely.[5] This trajectory, understood by Negroponte while he studied how architects might someday use computers, is legible in the transition from CAD to BIM, although the current use of BIM is hardly disruptive, yet. While adoption of BIM is widespread in western building economies[6] it continues to be used in the service of traditional project methodologies, including drawing-based design delineation and typical delivery methods and its full capabilities in the parametric are only now beginning to emerge as a potentially disruptive influence, enabling new capabilities in project collaboration, simulation, digital fabrication, and facilities management.

But BIM as both tool and methodology is far from the "standard" that CAD has become. The technology and accompanying design process changes are still evolving rapidly in parallel with changes in practice protocols necessary to fully deploy its benefits. In the near term, use of BIM is easier as computers have become faster, storage cheaper and more broadly accessible, and interconnectivity between machines deeper. Social networking methods are rapidly replacing traditional "store and manage" web sites as the infrastructure of collaboration. The last generation of CAD drafters is fading from the scene to be replaced by their more model-savvy successors, the architecture and engineering graduates of the last decade. As challenging as members of both generations have found the transition to BIM to be, the real disruption predicted by Negroponte is the advent of parametrics in combination with the accelerating capabilities of BIM empowered by cloud computation.

Today's generation of designers take the availability and utility of modeling for granted. The widespread use of scripted geometry modelers—primarily Rhino driven by Grasshopper—has spawned a set of young designers with different skills and sensibilities from their analog (or even CAD-enabled) predecessors. Design representation today is strongly rooted in digital models and scripts, and the generation of form/idea comes in equal parts from direct manipulation of geometry as the output of algorithms. A script represents both a strategy for a solution and the imbedded logic of its intent. When that set of sensibilities and methods is applied to the representational potency of cloud-enabled BIM, Negroponte's anticipated disruption is inevitable. The "things that architects make" will be different, the ways we create them new, and the reasons that we make them will change. The results must inevitably be different as well.

For these purposes, we will posit the rapid and eventually complete evolution of two technologic components: BIM as a platform for ideation, representation, and simulation; and parametrics as a method for accessing, interrogating, and manipulating the characteristics of that BIM-based representation. The modeler instantiates the representation of the building in digital form, and access to its characteristics (and their subsequent manipulation by scripts) provides the parametric opportunity. We will further stipulate that the computational infrastructure that makes both possible is shifting from desktop computers to cloud computing, unbinding parametricism from the limits of computational power (CPU resources) or data storage and access. You can create and save as much information as you want, move it virtually anywhere, access any version on multiple devices from studio to job site, and apply essentially infinite processing cycles to that information.

Irrespective of the ends to which these powerful capabilities may be applied (which is the subject of much of the balance of this book), it is clear that much broader portions of the design enterprise will be robustly represented, simulated with accuracy, and analyzed extensively. Drawings and charts are replaced by digital models and analysis software. Models are more detailed, and evaluation of those models by algorithms will run in the cloud simultaneously with design manipulation. Decision-making based on intuition, rules of thumb, or outsourced engineering talent will be replaced with immediate feedback on the technical implications of a given design strategy. Where a design was once represented almost exclusively through the proxy of geometry projected on a paper drawing, higher resolution representation will include not just more detail, but performance, materiality, even imbedded construction logic. Imbuing representation with parametric formulation extends that representation with both intent and result.

This is a critically important if largely underestimated change in design methodology that is bubbling up from the workstations of today's digitally empowered designers, with implications for the nature of the design process and the role of architects themselves. The traditional patriarchal relationship between architect and engineer (and, to a lesser extent, architect and builder) is potentially toppled by the emergence of computation analysis because work that was once demanded of engineers during the development of a design ("make this stand up" or "keep this heated or cooled" or "operate this within the following energy load") can be done, at least in part, by analytical software running in parallel with modeling software. Provisional answers to questions about what may or may not work are immediately available to the architect running a model and, for example, energy analysis or code simulation tools simultaneously. More rapid generation of alternatives, culling of options that fail to meet certain performance criteria, and a general assignation of technical evaluation to computation should result in better design, at least in theory.

Computers have long been used for analytical purposes in architecture, through purpose-built software assigned to tasks like energy analysis or structural evaluation, or more generic strategies empowered by the ubiquity of spreadsheets. Parameters, by definition, bound the solution space of analytical results. But the shift described here is in the relationship of these tools to the designs themselves when directly connected via those parameters to the underlying digital design model, and the potential to use optimization strategies like scripting to manipulate those models based on the results of analysis. A parametric strategy could be used in such a way to precisely locate a building on its site relative to its openings, glazing materials, and movement of the sun. Automating the process to answer that question without direct human intervention frees that human to spend her time solving more ineffable questions of composition—or drinking coffee.

Techniques already in use in manufacturing such as multivariable design optimization[7] use computational means to "pre-select" ranges of design options based on the parameters of performance and outcome defined by the designer at the outset of the design exercise. These methods create the "exploration space" within which a range of solutions can be found and selected, largely because the analytical models that underlie the design are explicit, accessible, and available for iterative evaluation. When computation is infinitely available and largely free, analysis is no longer a burden to be outsourced to engineers but a design resource in and of itself.

When design heuristics and analysis become closely bound together, the fundamental methodologies of design are radically altered. The definition of the problem space—the range of possible viable solutions—can be generated from underlying intent that has been defined by the selection and manipulation of key project parameters. The generation of possible solutions is unconstrained by computation cycles, computerized or human, and thus broad iteration of answers is possible. Those answers can be evaluated and measured by simulation algorithms—computer programs that operate alongside those that generate answers—for informed selection. The mathematics of multiple analytical models that are the basis of ideation and iteration is explicit, accessible, and usable in support of finding reliable and predictable results. Analytical models now join their physical and digital counterparts in the architect's toolkit. And, "parametricism" per se is redefined to include performative optimization toward results.

While these ideas are interesting in and of themselves as a strategy for improving the much maligned efficiency and effectiveness of the building industry[8] they have far greater implications for the role of architects, the means by which they work, and industry processes and models in which they must operate. Design projects that are extensively simulated and evaluated have measurable characteristics. Those measurements can be the basis of predictions of building behavior and performance, constrained only by what we can reasonably represent and evaluate in simulation. And reliable predictions become the basis for commitments upon which the underlying value propositions of building can be reformed.

A prosaic but illustrative example can be found in the straightforward obligation for building water-tightness. Setting aside for the moment the ineffable qualities of a well-designed building (however you may choose to define them), it can be reasonably expected that a building designed by a professional architect and constructed by a competent builder will not leak. But it often does; opened to the watery elements by some mixture of inadequate design, poor materials, shoddy workmanship, unanticipated weather conditions, or an inattention to proper building maintenance. A typical building project has no obvious or immediate line of accountability for this issue as the problem may have been generated in equal part by the designer, the builder, and the owner, in concert with the elements. A poorly conceived flashing detail, some mis-installed caulking, or a failure to properly maintain clear roof gutters are all equally potential culprits. And, no one takes full responsibility for this fundamental aspect of building performance precisely because the interaction of various obligations and processes controlled by various players makes results unpredictable and promises impossible to make. When a failure does occur, diagnosis is difficult and blame often assigned during litigation.

But imagine if a building's ability to repel water was predictable by virtue of careful analysis and simulation of the building's choice of materials, the logic of its construction, its thermodynamic behavior, and its owner's use. A combination of parametric definition ("what do we want this building to do?"), evaluative simulation ("how will it go together and then operate?"), and detailed knowledge of materials and similar installations ("what has happened in other like circumstances?") could lead an architect to make a promise that the building would not leak. Such a promise, at odds with every generally accepted strategy for risk management today, would increase the architect's credibility, maintain attention rightfully on the architect's more ineffable intentions, and likely result in higher value (and, therefore, higher compensation) to the architect herself. This methodology applied to larger questions, like the economic effect of a building or its impact on the environment writ large, is an opportunity for architecture itself to expand its breadth, reach, and importance.

As sensor technology and reality-capture techniques like laser scanning become more embedded in the building operations process, designers can further access huge amounts of data about how other buildings are performing after design. Comparing those results, and evaluating design options based on that evidence, creates another "big data" stream of input into the ideation process, a basis of reference for precedent and outcomes, and assurance that a promise of a result is backed up with real-world experience, all potentially supported by parametric analysis.

The evolution of design computation suggests that the technologic capability to work in this manner will soon be possible, constrained only by the willingness of architects to embrace a new paradigm of obligation and service. But the track record of the profession is not promising in this regard, in that architects have spent most of the twentieth century disabusing themselves of responsibility for many aspects of buildings, and particularly the operating results of their efforts. The oft-heard cry of "loss of influence and control" has been matched by a disinterest in taking responsibility for the measurable outcomes of building, particularly building performance, construction cost, and schedule. A class of professionals willing to do so has emerged accordingly.

Other important parts of the modern economy are embracing outcome-based methods in relentless efforts of self-improvement: data mining for drug outcomes, large-scale climate modeling and analysis for crop yields, digital simulation as the basis for automobile design, manufacturing and marketing, and virtualization of film and gaming. The medical industry is experimenting with compensation models tied to patient outcomes rather than number of procedures performed.[9] Those trends set expectations for the building industry and circumscribe its potential to reform its objectives, processes, and result. Architects can and must position themselves in the design and construction process anew.

3

Modern construction economic systems—that define the ways architects are involved and paid, and the means of procuring and executing building—derive from a basic exchange of financial value between the participants and the underlying assumptions about the related value of each player's contribution. These systems today are based largely on simplistic, reductive, and mostly ineffective strategies designed to limit short-term exposure to cost and risk, and are rarely deeply connected in any way to the underlying intent of the building projects they deliver. In fact, the most common performance metric in most delivery models is "lowest first cost"—choices are almost always made based on the lowest proposed cost of any component in the delivery system, be it architectural fees, contractor bids, or material prices. The contributions of the players are thereby undervalued by definition even when projects begin with the highest social aspirations. While the reasons for this situation are complex, there is increasing dissatisfaction with results among building clients who regularly experience cost overruns, schedule misses, and technical failures. As recent innovations in project delivery suggest, they are looking for another way.

The introduction of BIM to the mainstream building market spurred what may now be seen as the first move away from commoditized building economics toward outcome-based models. Rapidly evolving systems known generally as "integrated project delivery"[10] create teams of co-dependent designers, builders, and owners who share the risk and reward of building projects that are measured by mutually agreed outcomes that benefit the project rather than individual contributors to that project. The informational transparency of BIM created the opportunity from which integrated projects now spring, but the ability of a project team to understand outcomes based on a deeper understanding of a digitally simulated parametric project could be the most important shift. Broader more powerful parametric simulation means deeper more important outcomes can be predicted. The underlying means of production will shift accordingly and the value created and delivered by architects dramatically improved. And as analysis algorithms and approaches move from the more proximate technical problems of building performance to broader sociological or behavioral simulation, the architect's ability to project outcomes, promise results, and create real value improve significantly.

Further, the fundamental legal standard by which architectural competence is measured, the standard of professional care, by its very definition inhibits innovation and deep process change.[11] Because it defines proper professional judgment in terms of what another competent practitioner might have done in the same circumstance, pushing methods and design techniques forward in the manners that are described here is not just difficult, it is professionally dangerous. And, a generally accepted principle of risk assumption that is vigorously enforced by providers of architect's professional liability insurance is the avoidance of guarantees or warranties; provision of professional services means never promising actual results. So the standards of risk, and with them the architect's willingness to embrace it, must be disrupted; speculations of similar competence replaced with measurable results based on commitments made early by the designer. Those commitments could be based on the results of computational, parametric simulation of the building considered and eliminate what is possibly the single greatest barrier between architects and their potential to create measurable results.

Larry Ellison's success with his catamaran was, in some ways, predictable. He was confident that the right combination of technological prowess—design simulation that leveraged operational parametric telemetry—would change the odds of success in his favor. With formidable technology (and a lot of cash) he disrupted the age-old sport of sailing. That success was measured by the speed and agility of a sailboat on the open water against a formidable opponent and made possible by the control of the physical by the digital. As architects we can set the definition of success in design and building as we see fit, if we are willing to deploy the available tools and provocative methodologies that result. It is an opportunity not to be missed.

Notes

1. Mario Carpo, *The Alphabet and the Algorithm* (Cambridge, MA: MIT Press, Cambridge, 2011).
2. As further described by Carpo in *The Alphabet and the Algorithm*.
3. Moore's Law, invented by Intel's Andrew Moore early in the age of personal computers, suggested that the overall processing power of computers doubles every two years.
4. See, "Archaeology of the Digital," at http://www.cca.qc.ca/en/exhibitions/1964-archaeology-of-the-digital [accessed August 15, 2014].
5. Nicolas Negroponte, *The Architecture Machine* (Cambridge, MA: MIT Press, 1970).
6. BIM adoption in the United States has crossed 70 percent among architects, engineers, and builders, and European adoption is greater than 30 percent, according to McGraw Hill Construction Analytics, as described in their report *Smart Market Report: The Business Value of BIM* (New York: McGraw-Hill Publications, 2009).
7. Wikipedia defines multidisciplinary optimization as follows: "Multi-disciplinary design optimization (MDO) is a field of engineering that uses optimization methods to solve design problems incorporating a number of disciplines. It is also known as multidisciplinary optimization and multidisciplinary system design optimization (MSDO). MDO allows designers to incorporate all relevant disciplines simultaneously. The optimum of the simultaneous problem is superior to the design found by optimizing each discipline sequentially, since it can exploit the interactions between the disciplines. However, including all disciplines simultaneously significantly increases the complexity of the problem." See http://en.wikipedia.org/wiki/Multidisciplinary_design_optimization [accessed July 21, 2014].

8. Worldwide construction is largely considered to be inefficient, with waste factors as high as 40 percent. For further information, see "Rethinking Construction: Report of the Construction Industry Task Force," at http://www.constructingexcellence.org.uk/pdf/rethinking%20construction/rethinking_construction_report.pdf [accessed November 1, 2014], especially p. 15, item #25.
9. See "Health Insurers Are Trying New Payment Models, Study Shows," in *The New York Times*, September 7, 2014, or online at http://www.nytimes.com/2014/07/10/business/health-insurers-are-trying-new-payment-models-study-shows.html?emc=eta1 [accessed September 8, 2014].
10. For early definitions of integrated delivery, see the AIA's (American Institute of Architects) "Integrated Delivery: A Guide," at http://www.aia.org/contractdocs/aias077630 [accessed September 16, 2014].
11. Professional competence (or its lack, negligence) is defined by positing "what would a competent professional in similar circumstances have done?" This standard is largely a matter of judgment, and the definition of "competence" and "similar circumstances" broad and undefined.

Chapter 14

Spinoza's geometric and ecological ratios
Peg Rawes

This chapter questions the relationship between the latest evolution of digital architecture, parametricism, and the claim that these computationally driven design methods also establish new environmental conceptualizations for architecture.* Continuously increasing rates in climate change accentuate the need for the profession to responsibly develop sustainable design approaches for tackling the attendant social and environmental issues of resource-depletion and pollution. Within this context, the digital design community has developed a vocal set of agendas which, it argues, align its manipulation of computational software with wider notions of organic and biological life and often, by implication, "environmental" awareness. However, substantial climate change evidence also shows that economic and social action is required, rather than another design "style" driven by modernist principles of autonomy. I therefore ask if these computational software geometries can be considered ecological in any viable way; and, if there is to be any conversation about process-based, materialist architectures, just who benefits from these cultures? Is parametricism of any real value for aiding the design of non-pathological, ecologically responsible buildings, which require ratios of ethical, economic and material intelligence, rather than myths of emergence?

I question the seductive rhetoric of parametric discourses, asking if their digital geometric techniques enable diversity on a number of fronts: first, can they

---

*An earlier shorter version of this chapter was given as a keynote address at "Interstices Under Construction Symposium: Technics, Memory and the Architecture of History," at the University of Tasmania, Launceston, Tasmania, in November 2011, and published as "Spinoza's Geometric Ecologies" in *Interstices 13: Journal of Architecture and the Related Arts* (Auckland: Enigma: He Aupiki, 2012), pp. 60–69. We would like to acknowledge and thank the publisher Enigma: He Aupiki, Auckland, New Zealand, for their generous permission in allowing us to publish a significantly revised and updated version of Peg Rawes' essay.

address the need for designing cities, buildings and infrastructure which, in the best case scenario, contribute to nurturing ecological biodiversity, or at the very least, do not contribute further to carbon emissions and neoliberal ignorance of climate change (agnotology)?[1] Second, how exactly do digital parametric geometries enable the necessary ethics of wellbeing that global and local communities require, both in society as a whole, but also in the architectural profession? For example, why is parametricism promoted so strongly by its proponents as an innovation of "nonhuman" authorial subjects, rather than focusing their efforts on changing the unequal economic and social ratios that can improve diversity in the profession's employment practices and beneficiaries, underpinned by structural change?

I develop my argument through materialist and biopolitical philosophical thought, drawing from an early-modern philosophy of "wellbeing" and contemporary feminist and biopolitical inquiries into subjectivity, in order to explore the relations between geometric thinking and human-nature relations. Baruch Spinoza's "geometric thinking" in the *Ethics* (1677)[2] provides a fascinating conceptualization of an unusually ecological and relational form of rational thought, developed in his concept of *ratio*.[3] This seventeenth-century text therefore predates parametricism's self-conscious, romanticized discussions of origination in Darwinian and early twentieth-century biological morphologies by two hundred years. It is distinct from these later biological discussions of evolution because Spinoza's anti-Enlightenment thinking raises important and searching questions about the social and the political geometric ratios in "design" which are necessary for today's ecologically-responsible architectural design. Instead of continuing the reliance upon outdated narratives of style that are predicated upon reductive epistemologies of "self-same" replication, Spinoza's ratio contributes to the debate because his concept is formed out of *dissimilarity*.

Parametric designers argue that the new "relational" geometries, which software such as Rhino and Grasshopper deliver, enable complex and self-evolving algorithmic or "biological" morphologies to be generated. Prominent in the parametric debate, and present in this volume, Patrik Schumacher has written that it:

> finally offers a credible, sustainable answer to the drawn out crisis of modernism that resulted in 25 years of stylistic searching. (...) As a conceptual definition of parametricism one might offer the following formula: Parametricism implies that all architectural elements and complexes are parametrically malleable. This implies a fundamental ontological shift within the basic, constituent elements of architecture. Instead of the classical and modern reliance on ideal (hermetic, rigid) geometrical figures - straight lines, rectangles, as well

as cubes, cylinders, pyramids, and (semi-) spheres - the new primitives of parametricism are animate (dynamic, adaptive, interactive) geometrical entities - splines, nurbs, and subdivs - as fundamental geometrical building blocks for dynamical systems like "hair," "cloth," "blobs," and "metaballs," etc. that react to "attractors" and that can be made to resonate with each other via scripts.[4]

However, in attempting to define a system of formalist aesthetic criteria, Schumacher also repeats the familiar modernist myths of autonomy and innovation. Presented within a more explicitly biological framing, Michael Hensel and Achim Menges's computational "morpho-ecological" method has developed a "framework (…) firmly rooted within a biological paradigm (…) [and] issues of higher functionality and performance capacity,"[5] in which the building and its envelope are improved through morphological adjustments within the system (and so recalls Christina Cogdell's critique of "fitness" in her chapter in this volume). Susannah Hagan's urban form of ecological parametricism suggests a design method that is more reflexive between the digital, the social and the site specific, when she writes: "Environmental metrics can be used to generate parametrics. Parametrics are now firmly embedded in experimental design." However, in a manner very similar to Hensel and Menges, she still prioritizes geometric thinking as a principle of morphological "form-finding and the relationship between form and performance in the interests of a new and elegant economy of means"[6] In *The Sympathy of Things*,[7] Lars Spuybroek's focus on Ruskin's naturalism and the genetic principles of emergence in Darwin's theory of species provides an unabashedly romantic account of nature, technology and biological evolution in architecture.

Taking a less idealized view of parametricism's capacity to reproduce nature through computational form, Mario Carpo's genealogy of digital architecture notes the underlying similarity of these supposedly diverse projects when he defines parametricism as a "function which may determine an infinite variety of objects, all different (one for each set of parameters) *yet all similar (as the underlying function is the same for all)*."[8] More critically, in "Architecture and Mathematics: Between hubris and restraint,"[9] Antoine Picon points to the historical shift from classical principles of geometry to modern mathematical forms of calculus in eighteenth-century European architectural design. Following Gottfried Wilhelm von Leibniz's and Girard Desargues's respective innovations in calculus and projective geometry, he observes that technical advancements in architectural geometries fundamentally changed power relations between nature, technology and generative design principles. After calculus, Picon notes, design institutes both the potential for "unfettered" invention and "hubris"; a loss of "restraint" which, he suggests, bears a resemblance to the contemporary issue of sustainability. He asks if we need a return to mathematic "restraint" in the face of pressing questions about resource depletion, and the need for architectural design that is not primarily determined by the perception that its "power" is located in principles of autonomous genetic digital production:

> To conclude on this point, one may observe that this polarity, or rather this balance, has been compromised today. For the mathematical procedures architects have to deal with, from calculus to algorithms, are decidedly on the side of power. Nature has replaced God, emergence the traditional process of creation, but its power expressed in mathematical terms conveys the same exhilaration, the same risk of unchecked hubris as in prior times. What we might want to recover is the possibility for mathematics to be also about restraint, about stepping aside in front of the power at work in the universe.
>
> It is interesting to note how the quest for restraint echoes some of our present concerns with sustainability. The only thing that should probably not be forgotten is that just like the use of mathematics, sustainability is necessarily dual; it is as much about power as about restraint.[10]

Picon's scepticism of the supposed freedom that is attributed to modern computational forms of geometric invention reflects my discussion about whether the power invested in these new digital processes is effective, or even desirable, for tackling the complex issues of social, economic and environmental inequalities in the built environment that are linked to climate change, now visible in the Global North, as well as in the Global South, including; restrictions to safe and sustainable water and energy provision, high quality affordable housing, public services and healthcare.

It is also worth noting the professional and training contexts in which parametricism is situated, especially when questions about how diversity or difference are actually structured into the profession remain critical: when mathematics and computation are presented to students yet again as the "new" universal grammar and language of innovation (this isn't the first time that zeros and ones have seduced designers); when commercial design companies are increasingly being driven by large-scale digital data systems, including BIM, which tend towards the normative and economically 'best fitting'; when women make up fifty percent of the students who train in the discipline in the UK/US, yet this is a branch of architectural design more commonly populated by male students who are instructed through intensive workshops which focus on repetition and emulation. Commonly, the most successful students then progress from these workshops to join hermetic teams or "labs," but do not necessarily benefit either from the financial returns of their work, or the recognition of their authorial contribution.

This "hot-house" training of designers cultivates parametricism's metaphorical alignments between digital code and biological DNA as evidence that architects have now been released from the limitations of human authorship in design processes.[11] In addition, its supposedly bottom-up non-anthropomorphic design origins are commonly aligned with the materialist philosophies of Manuel DeLanda, and the speculative realist philosophies of Quentin Meillassoux and Graham Harman, who critique anthropocentric thinking and propose a return to objective "facticity."[12] For parametric architects, the aim of such associations is to show that architectural design is further embedded in non-ontological meaning; i.e. nonhuman production. However, since nurturing social responsibility, agency and environmental relations in the built environment are clearly still necessary design intelligences required for tackling the complexity of ecological relationships between human-made nature (i.e. the built environment) and nonhuman nature (i.e. the natural environment), whether these philosophers' exclusive rejection of situated ontological practices is most sufficient, socially ethical or environmentally informed, is questionable. Rather, these thinkers can be charged with perpetuating a logic that *still fails* to address other *indisputable* non-anthropomorphic realities (facts), such as the facticity of capital in architectural design markets, or the depletion of non-replaceable material resources. (Digital architects also forget or ignore the long aesthetic and materialist philosophical history that has explored the human-nature versus nonhuman nature conundrum since the seventeenth century, including Kant's theory of the sublime, and Marx's theorization of human-nature relations.)

Outside these narratives, however, feminist philosophers have *already* convincingly shown that complex material and social relations – including sexual, biological, social and political difference – clearly do exist not only in art and culture, but also in science and technology, when they are used responsibly for nonpathological ends. These discussions provide transformative understandings – available to women *and* men – about how ethical power relations between human and nonhuman realms enable more sophisticated discussions about the status of design agendas that claim to be non-anthropomorphic. Their definitions of positive sexed biological and cultural differences (again, what I am calling "ratios") are necessary for parametricism, precisely because it claims that the drivers for non-standard computational design are located in organic processes and biological forms – most commonly present in the rhetoric that digital code and scripts are computational "DNA" that "originates" new complex morphological design. As I have indicated above, such claims expose parametricism's dependency upon the replication of self-similar forms (even if they are topological rather than geometric). In fact, this uncritical belief in computationally-driven emergence reduces diverse organic life to a self-same matter: for example in their over-determined return to historical sources such as D'Arcy Wentworth Thompson's classical study of evolutionary biological morphology, On Growth and Form,[13] rather than understandings of contemporary biophysical sciences, such as epigenetics – again put forward in this volume by Christina Cogdell – which, like sexed understandings of society and environment, are more closely aligned to coterminous and non-teleological processual relations, rather than repetitions of the same. Most discussions about digital morphologies, even if they emphasize an emergent principle of production, also fail to distance themselves from the vicissitudes of form/matter distinctions, even when they are non-standard geometries or topologies or when claiming the sublimation of form to matter, because they reuse weak notions of both terms. Omitted throughout is a concern with structural change to geometric thinking in which ratios are driven by economic, ecological and nonhuman ontologies.[14]

Focusing now on Spinoza's seventeenth-century conceptualization of geometric thinking, subjectivity and environmental relations, clearly it is historically distinct from our advanced capitalist, technocratic and global contexts. But his thinking is valuable for a number of reasons; first, because his theory of "subjectivity" does not accord with the moral autonomous modern subject; secondly, because his principle of nature provides an original non-anthropocentric mode of rational thinking about human and other relations; and, third, because he was writing during a period of intense mercantile and social change in Amsterdam, and the Low Countries, providing a context through which I consider social and economical geometric ratios which define the seriously dysfunctional UK housing crisis.

...

Spinoza's "ratio" is an unusually ecological form of geometric thinking. It is a highly complex early modern expression of natural biodiversity, where commonality between human and nonhuman entities exists, but as expressions of dissimilarity or disequilibrium, rather than an evolution or inheritance of similarity. Spinoza's principles of self-determined and co-existing entities are not "designed" to anthropomorphically reflect the author (i.e. the subject) in the post-Enlightenment sense. Neither are these entities simply reducible to metaphors of universal morphologies, genetic codes or algorithms. Rather, they are expressions of geometric and ecological relations. This relationality means that the subject and nature are constantly understood to be in process, what we might call "subjects-in-process" or non-teleological ratios. Moreover, because Spinoza's geometric ratio pre-dates Enlightenment's moral subject or the determinism of twentieth-century biological replication – its nonhuman tendencies are more diverse. We might say that it is a nonhuman "ecological geometry" in which nature is not determined into simple units by the idealized and hermetic application of computational scripts, which promote "unfettered" generativity for neoliberal ends.

Two key ideas compose Spinoza's geometric or ecological ratio: first, his notion of nature, which he also calls "substance"; and second, his concept of self-determining agency or life, "conatus." In Definition 3, Spinoza writes: "By substance I mean that which is in itself and is conceived through itself; that is, that the conception of which does not require the conception of another thing from which it has to be formed,"[15] and in Part 3, when expressed as the conatus it is "the power of any thing (…) by which it endeavours to persist in its own being."[16] Together, these life forces form the basis for a non-anthropocentric power in his thinking. Substance is a complex term that is not reducible merely to inert matter or form because it designates extended and potential existence. It constitutes the immanent biodiversity of life in Nature – in humans and animals, in non-organic entities such as stones and geometric ideas, and also in the "divine" relation God-as-Nature. And, in conjunction with Spinoza's other wonderfully complex idea of life, "expression," Substance/Nature constitutes an immanent plenitude of possible and actual realities at different scales and in different modes: from the divine to the everyday, from the absolutely non-anthropological entity, to the individual's corporeal expression of joy or distress. Spinoza tells us that there is also a ratio between two forms of this productive power in Substance; on the one hand, its potential for change (*Natura Naturans*), and, secondly, the manner in which it creates diverse modes of existence (*Natura Naturata*).[17] When considered together, this ratio of natural powers therefore generates a plenitude of ideas, bodies and entities in the world so that Spinoza considers substance to be a "univocal" concept of life in all its material (e.g. biological) and immaterial (e.g. psychic) manifestations; including also, importantly for the discussion here, architectural and geometric processes. We might say that in this "natural" architecture, all modes of life – human and nonhuman – are ratios of substance. In contrast to contemporary mathematical and geometric methods that exclude "natural" processes from the pure reason of disembodied/non-ontological procedures, Spinoza's geometric thinking is firmly constituted in sensory realms *as well as rational* relations.

As has been well documented in Spinoza scholarship, in Part IV of the *Ethics*, with meticulous detail, Spinoza explains how ordinary people express these constantly changing ratios in their daily desires and fears, and in everyday common ideas.[18] Here, ratio is an intensive series of incremental "affects" that form processual subjectivity and an account of dynamic human and nonhuman wellbeing. He identifies how the ratio between active and passive expressions exists in many modes, including: cheerfulness, pleasure, titillation, hatred, fear, anger, contempt, disparagement or melancholy. It is an account of human-nature which accords not with the Enlightenment traditions of the moral autonomous subject, but affirms diverse, continuously changing powers in the subject. Taken in our contemporary biopolitical context we can suggest that these capacities constitute the dissimilarity and non-equivalence of self-determining subjects, and which biopolitical and feminist philosophies locate in the *other* notion of life, *zoe*, rather than *bios*.

Drawing partly from Michel Foucault's 1970s lectures on biopolitics, Donna Haraway, Rosi Braidotti and Clare Colebrook's feminist ecologies present nonhuman ontologies which do not exclude economic and political "facticity" because relations are *always* situated in relation to questions of power, and following Foucault's critique of biopower, specifically, neoliberalist governance. These biopolitical ontologies show that self-same subject-centred thinking categorizes and manages human life (*bios*) through normative moral social forms, and excludes complex *other* nonhuman subjectivities (or *zoes*) which *also already* exist, but are not legitimized by neoliberal society. Seen from these perspectives, parametricism's claims for computationally originated universal design principles are exposed to be dependent upon idealized and neoliberal codes of production, which fail to take into account socio-economic and political ratios – including human labour and irresponsible uses of advanced technology.

Spinoza brings all entities, whether they are naturally occurring or human-oriented, into a special kind of biophysical process, in a manner that also previews important twentieth-century ecological and vitalist theories including, Arne Naess's "deep ecology," Gregory Bateson's "ecology of mind," and Deleuze and Guattari's "geophilosophy". Notably, this life principle is derived from the divine power of God/Nature, contrasting with computational architecture's scripts, which are described as self-organized "genetic" code. In Spinoza's "natural" geometry, biodiversity calibrates all entities at all scales, but not reducible to a simple digit or unit of computational code. As a result, he underscores that human subjects and geometric figures are manifestations of *nature-in-process*. This ecological notion of immanent differentiation produces absolutely unique expressions of substance: such as, the specificity of trees, stones, horses or dogs; or the difference between feelings of delight, disappointment, rage or fear; the capacity of the emotions to generate action and transformation, or the diversity of physical and psychic expression in architectural ideas or designs. Spinoza's geometric ecology is therefore always inherently concerned with *diverse living* relationships, not just formal or material self-same relations.

Substance also has a complex ecological meta-structure because it expresses a unique triad of relations between the three special geometric "elements" that Spinoza invents – "attributes," "modes" or "affects," and "common notions." These powerful transversal relations also generate an intense diversification into further geometric elements, such as, definitions, axioms, propositions, corollaries or scholia. The complex triadic ecology of relations between the attributes, affects and common notions is therefore an important historical preview of Félix Guattari's influential ecological thinking in *The Three Ecologies*.[19] Here, again, we have an example where the biopolitical nonhuman ratio of *zoe* is developed, for example, when Guattari defines it as *ecosophy*: "an ethico-political articulate (…) between the three ecological registers (the environment, social relations and human subjectivity)."[20] Guattari shares with Haraway and Braidotti an understanding that *zoe* enables positive expressions of self-determined life. For these thinkers, nonhuman subjectivities are distinct from normative ratios of biological life (*bios*) – now ubiquitous at the scale of biometric management of the self in society through social media – or the negative categorization of "bare life," in which the nonhuman subject is excluded from social norms and managed through governance by violence and fear (see the important biopolitical critiques of the *inhumane* developed by Giorgio Agamben, Judith Butler and Jacques Rancière). Rather, in Spinoza's psychophysical affects or ratios we find affirmative self-determining nonhumanist alterities. In each case, ratio is directed towards "self-determination" or "care" of the self, which accords with ethical and non-pathological relations that current posthumanist feminist and biopolitical ecological thought value.

Biopolitical geometric ratios are therefore not constructed by disembodied transcendental laws of reason, or digital metaphors of biological processes, but out of the everyday and transformative expressions of the body within its own singular environment or habitat. Consequently, we might also say that Spinoza's attention to the habitus of the subject-in-process previews Haeckel's 1866 definition of the science of ecology that observes the entity in its home, habits, habitat or milieu; or Jacob von Uexküll's theory of *umwelt* (1909), a "biological semiotics" through which he theorized the coterminous existence of the organism with its specialist habitat.[21] In addition, my attention to the ratios in complex processual *corporeal* and technical historicity of geometric expression acknowledges previous architectural historical studies of geometric thinking and practices, including Picon's writing and Alberto Pérez Gómez's *Architecture and the Crisis of Modern Science*,[22] and, closer to my discussion, Robin Evans's recognition of everyday geometric figurations in *Translations from Drawing to Building and Other Essays*.[23] But Spinoza's project is also distinct from Evans's because it is concerned with the power of nonhuman differentiation.

...

Antoine Picon's critique of unfettered digital design also connects with my concern that parametricism repeats the long tradition of disembodied, neutral or "unsexed" reason. His argument helps me to open up the space and to ask if "an-other" ecology of geometric relations is possible. However, whilst his critique of the relationship between geometry, proportion, God/Nature and sustainability certainly reflects key principles in Spinoza's geometric ecology, Spinoza's notion of immanent relations are distinguished from the later technical rationalization of geometric thought, in particular, because ratio is more clearly determined by a principle of biological *duration*.

Spinoza explains how the power of the "conatus," immanent in all human endeavours, generates geometric, aesthetic and architectural modes of expression. Unlike the Enlightenment's deployment of calculus, it is not a subsumption of substance's power to a principle of unfettered infinitude *or* vitalism, nor is ratio instrumentalized by reductive anthropomorphic knowledge. Instead, Spinoza's elaborate examination of the ratios between his "elements" – the attributes, affects and common notions – shifts geometry from a disembodied procedure of logical deduction into a tripartite ecology that generates biodiversity within biopowers at multiple scales of differentiation and duration: for example, affects are transitive powers in themselves, which also express the attributes of mind and body. And, when the mind and body are most in "agreement" in the form of the "common notions" this ratio expresses an ecology that I have previously called "sense-reason" but in which the various biopowers remain differentiated.[24] As such, these durational powers constitute the continuously transitive subject-in-process that is essential for an ethics of geometric ecology.

Ratio is not just an "unfettered" principle of plenitude, but is ecological *because* its relations are always durational. Also, importantly, when it is expressed in the common notions, a geometric biodiversity is accessible *to all*, paradoxically because the "commonality" comes from its dissimilarity of ideas, bodies or geometries. Spinoza's natural geometric ratios are therefore less idealized biodiverse ecologies and lives; and, again, because of the differentiation with the principle of ratio, this biodiversity is not just a neutral or value-free materialism, but is biopolitical. As a result, Spinoza's thinking transforms the individuated experience into a complex social form of "communal" power, which is not idealistic, unfettered anthropomorphic progress, but accords more with contemporary philosophers' inquiries into ethical ecological thinking for the *uncommon* (*zoe*), and not just for those who can access these values through the market.

...

As indicated earlier, recent feminist philosophy promotes the ethical and nonpathological rights to aesthetic, political and material self-determination for *all* sexes.[25] In parametric literature there are moments of acknowledging sexual difference, but when this occurs it is the classically normative category of scientific biological material (*bios*): for example, Carpo cites Greg Lynn's introduction to *Folding in Architecture* (2004): "from the identical asexual reproduction of simple machines to the differential sexual reproduction of intimate machines"[26]; and Spuybroek acknowledges sex difference in Ruskin's critique of Darwin's theory of evolution, but restricts it to a discussion of beauty.[27] Thus, despite these brief discussions there is little evidence that it has been actualized as a self-determining agency by digital architecture's practitioners, cultures or products (although Beatriz Preciado's article on biopolitical "disobedience" in Francois Roche's issue on emergence is a small intervention).[28] It is also worth remembering that Arne Naess identified Spinoza's work as a precursor to his "deep ecology," especially for understanding the interdependent complexity of human, natural and built relations without recourse to instrumental or human-centred concepts of life:

> The specific thing to be learned from Spinoza (…) is, however, to integrate the value priorities themselves in the world. (…) Spinoza was heavily influenced by mechanical models of matter, but he did not extend them to cover "reality" His reality was neither mechanical, value-neutral, nor value-empty.
> This cleavage into two worlds (…) [of facts and values] can theoretically be overcome by placing, as Spinoza does, joys and other so-called subjective phenomena into a unified total field of realities.[29]

Spinoza's theory of Substance/Nature therefore generates not only absolute biodiversity or alterity in all beings (whether they be women, men, animals, trees, stones, geometric figures, etc.), but in his commitment to a "deep" rather than "shallow" value-specific biodiversity, which can be interpreted as a kind of proto-sexed theory of difference.[30] Feminist philosophers, Moira Gatens, Genevieve Lloyd and Rosi Braidotti have previously explored how Spinoza's affirmation of otherness is indeed a precursor to sex difference.[31] As such, its political and materialist biodiversity is a valuable historical example where ecologies or ratios of geometry, sense, reason and sex are reconfigured, and which may have valuable consequences for architectural design today that engages in geometric thinking. In the *Ethics*, then, geometric ecology might even be a sexed "technicity"[32]: its psychic and biological modes of differentiation constitute a special kind of rationalism for generating ecological biodiversity *in* the individual, society, the environment, and *in* contemporary architectural design processes. Spinoza's thinking resists the reduction of difference to simple human-centred (i.e. anthropomorphic) or instrumental understandings of nature and otherness. His affirmation of irreducible difference constitutes all entities, human and other, even if he does not explicitly describe or identify these as sexed (i.e. *not* gender-neutral) differences. More recent feminist philosophy that develops this sensibility in *critical* anlayses of advanced technology, such as digital architectures, includes Donna Haraway's "sympathetic critiques" of advanced technologies,[33] Rosi Braidotti's digital "ethics of care"[34] Elizabeth Grosz's feminist analysis of Darwin's theory of sex difference,[35] and Lorraine Code's socio-biological ecological thinking.[36] However, if digital geometric practices continue to remain oblivious to political, material and immaterial realities, including sexed difference, their claims for innovation are, paradoxically, limited by weak concepts of production which are seriously out-of-date for the needs of *all* twenty-first century architects, and their societies' *umwelts*, right across the planet.[37]

Biodiverse sexed geometric ecologies are also significant because of the continuing split between matters of "reason" and technology, versus "sense" and subjectivity politics that define many debates about ecological architecture. If current script-based geometries continue to reinforce the neutral/value-free universalism of western thought, and ignore "other" modes of subjectivity that are not restricted to simple models of anthropomorphic nature, matter or life, they perpetuate the self-same identity of neutral architectural identities, processes or histories from which they claim to break. Without a conversation about biodiverse sexed geometry, histories and theories of biodiverse ecologies and technicities that embed real difference will continue to be ignored, and technological and ecological values will continue to be seen as at odds with each other.

This discussion also reflects my concern about the way in which feminist theories of relations still often oppose the possibility that sexed ecologies and technologies can exist together, relying upon the essentialist division between sense as "female" versus reason as an exclusively "male" concern, and consequently always negative rational thought. In this formulation, sexed ecologies are effectively consigned permanently to understand ecology as anti-reason (as in Irigaray's outright rejection of technology).[38] Without addressing these schisms, feminist architects (male and female) will continue to be consigned to a-technological realms, rather than offering alternative notions of sexed technicities. The issue of ubiquitous technology versus the political, self-directed agency of the subject in environmentally responsive architecture has also been clearly established since the UN's 1987 Brundtland Report prioritized economic sustainability, enabling the architectural marketplace to generate sustainable development through anthropomorphically-driven "shallow" or "instrumentalized" technological remediation. Yet feminist discussions of nature and architecture that continue to view technology as *always* damagingly instrumental or alienating to society also perpetuate this exclusive split. Thinkers such as Braidotti, Haraway and Grosz have offered more challenging accounts of sexed technologies and science that are of value to architects who are involved in cultivating biodiverse ecologies with others. In addition, Spinoza's thinking may help to redefine the relationship between technology and reason *for*, and *by*, sexed subjects, and to question the reliance that feminist ecological critique has placed upon the relationship between sense, sex and the environment, yet to the exclusion of sexed reason and technology from these debates.

…

Finally, bringing Spinoza's inventive seventeenth-century discussion forward to today's biopolitical crises of individual and community wellbeing, we find that his analysis of the ratios that compose human nature also provides a means to consider the social, economic and material ratios that constitute our physical, mental and environmental health. Ratios of wellbeing are significant design constituents in today's UK housing provision, and the focus of a project which philosopher, Beth Lord and I are currently undertaking titled *Equalities of Wellbeing*.[39] A particularly pernicious form of ratio currently defines the UK's dysfunctional housing market, which is in a critical state of disequilibrium, partly resulting from the systematic decline in house building since the late 1940s, and the reliance of the UK financial markets on income from speculative housing (to the extent that it has drawn out criticism from the Governor of Bank of England).[40] In addition, in the past twenty-five years, the sell-off of local authority social housing since the 1980s, combined with the deep inequality in economic support for private owners and landlords through cheap mortgages, and the reduction in social welfare for low-income tenants in private and public rental markets has deepened the crisis. Researchers specializing in inequality and housing issues (e.g. charities including, Shelter, the UK's leading charity for homelessness, and The Joseph Rowntree Foundation)[41] observe alarming increases in overcrowding and in physical and mental health issues related to damp and poor living space conditions, such as asthma and depression. And, increasingly, these are being linked to the toxic ratios of cheap financial mortgages and loans offered to landlords and the speculative housing market versus "affordable" housing priced at unaffordable market rates for those on low incomes;[42] the recent "bedroom tax" on social housing tenants who do not use "extra" bedrooms; and leases which favor the landlord's rights to market rental rates over the tenant's rights to a home.

Other inequality ratios in housing provision include: lack of quality in space and material housing needs (e.g. light, heating, good quality build materials, sustainability, storage); the lack of regulation in the rental market which now means the cost of a criminally small (a "semi-studio") studio flat in London can be £1,000 per month[43]; the inflated housing price market in London, resulting in couples earning over £70,000 who cannot afford to buy a home in a Zone 2 London borough; and the systematic undercutting of quality of space provision in new builds by speculative housing developers so that a family of three can expect to have a property that is smaller by a bedroom than recommended levels, and where in London only eighteen percent of housing stock met recommended space standards.[44] In addition, the UK has the smallest ratio of space requirements for housing in Europe and new homes have been getting smaller: in 2005, just 76m2 (rather than the average house in UK 85m2), and in comparison with Irish new homes 87.5m2 (15 percent), Dutch 115.5m2 (53 percent), and the Danish 137m2 (80 percent).[45]

Epidemiological, social and economic research by Richard G. Wilkinson and Kate Pickett in *The Spirit Level*[46] and their charity, The Equality Trust (together with other housing and social policy charities) show how these dissimilarities of poor income, housing, education and healthcare construct chronic ratios of wellbeing in the UK represents a severe contemporary biopolitical crisis.[47] Data from these organizations shows how a biopolitical management of subjectivity at a number of interrelated scales generates dissimilarity in social relations; from the unequal distribution of financial support through mortgage products, taxation and welfare support by neoliberal forms of governance, to the reduced capacity for positive self-determination of those communities who are most unequal or dissimilar from the "norms." Moreover, these biopolitical relations of "geometric wellbeing and housing" are now being felt much more strongly across society – from the individual home-owner or private tenant, to the low-income family or elderly couple in local authority housing.

In response, professionals in the building and housing industries, including, architects, policy makers in housing and the construction industry, local authorities and housing associations, planners and charities, are increasingly pressurizing the UK government's management of housing and wellbeing in a manner that reflects Spinoza's thinking about the dissimilar composition of individual and social relations.

For example, Spinoza's ratios of wellbeing accords with the work now being done to set legal minimum criteria for high quality, sustainable housing provision by professionals across the building industry by architects, planners, local authorities, housing associations and housing developers.[48] Recently, for example, this research has been brought together to form the UK government's Housing Standard Consultation 2013–14.[49] In addition, this consultation process draws strongly from previous research undertaken in the London Housing Consultation 2010,[50] which provides detailed criteria for housing in relation to diversity and "place" shared and private space, dwelling space standards and "climate change mitigation and adaptation" Importantly, however, the positive potential that these "ratios of wellbeing" can provide for enabling housing quality is not new. There are many successful twentieth-century examples of high quality housing in the UK and beyond, which have enabled individual and social wellbeing. In the UK these include: the disastrously short-lived 1961 Parker Morris Space Standards which resulted in the design and build of high living-space ratios in council housing, and is one of the leading precedents for today's standards,[51] or the now-respected, large-scale post-war local authority housing projects (such as, Park Hill in Sheffield, Bevin Court, Islington, or the Alton Estate in Roehampton, Greater London) in which designers and planners realized housing that provided good models of ratios of space and resources for the individual household, and the community's needs.

These examples of *relational* professional design approaches to the UK housing crisis therefore show how "geometric ratios of wellbeing" are implemented to construct ethical and *humane* space and environmental standards of design and build, which do not result from idealized architectural design methods, or from negative management of individual's needs by policy makers, but instead reflect a set of affective ratios of wellbeing for diverse needs in society. Spinoza's commitment to ecological ratios might therefore enable debates about geometric ecologies in the discipline, and consequently, for the societies, and the human and natural environments in which we live and work.

Notes

1. See Lorraine Code's discussion of "Agnotology," in *Relational Architectural Ecologies: Architecture, Nature and Subjectivity*, ed. Peg Rawes (London: Routledge, 2013).
2. Baruch Spinoza, *Ethics: Treatise on the Emendation of the Intellect and Selected Letters*, [1677], ed. S. Feldman (Indianapolis: Hackett Publishing, 1992).

3. Thomas Picketty's *Capital in the Twenty-First Century*, trans. Arthur Goldhammer, (Boston: Harvard University Press, 2014) is the leading economic analysis of ratio in today's neoliberal societies, paying considerable attention to an economic history of the ratio between property, land value and income.
4. Patrik Schumacher, "The Parametricist Epoch: Let the Style Wars Begin." *The Architects' Journal*, Vol. 231, No. 16 (May 2010).
5. Eds. Michael Hensel & Achim Menges, *Morpho-Ecologies* (London: AA Publications, 2006), p. 16.
6. In, eds. M. Mostafavi & G. Doherty, *Ecological Urbanism* (Baden: Lars Muller Publishers & Harvard, MA: Harvard University Graduate School of Design, 2010), p. 462.
7. Lars Spuybroek, *The Sympathy of Things: Ruskin and the Ecology of Design* (Rotterdam: nai010 publishers, 2012).
8. Mario Carpo, *The Alphabet and the Algorithm* (Cambridge, MA: MIT Press, 2011), p. 40; *my emphasis*.
9. Antoine Picon, "Architecture and Mathematics: Between Hubris and Restraint," in *Mathematics of Space: Architectural Design*, Architectural Design Profile Number 212, ed. G. Legendre (Chichester: John Wiley & Sons, 2011), pp. 29–35.
10. Ibid., p. 31.
11. Carpo also warns against overstating the radicality of agency in the new software (and Web 2.0) technologies, since these have clear authorial hierarchies embedded into them. See, Carpo, *The Alphabet and the Algorithm*, p. 126.
12. Sanford Kwinter refers to parametricism and these ideas as either "atrocious" or "very interesting." See, Sanford Kwinter, "Ecological Thinking," at *Proto/e/co/logics: Speculative Materialism in Architecture* seminar, Croatia, 2011, http://vimeo.com/28810672 [accessed April 2, 2012].
13. D'Arcy Wentworth Thompson (1915), *On Growth and Form* (Cambridge: Cambridge University Press; New York: Macmillan Press, 1945).
14. See, for example, Judith Butler's critique of form and matter in, Judith Butler, *Bodies that Matter* (New York: Routledge, 1993).
15. Spinoza, *Ethics*, p. 31.
16. Ibid., p. 108.
17. Ibid., pp. 51–52.
18. Ibid., pp. 156–95.
19. See, trans. I. Pindar & P. Sutton; Félix Guattari (1989), *The Three Ecologies* (London: Athlone Press, 2000).
20. Ibid., pp. 19–20.
21. For example, see (trans. Brian Massumi) Giles Deleuze & Félix Guattari, *A Thousand Plateaus: Capitalism and Schizophrenia* (New York: Continuum, 1996), p. 257; and, Elizabeth Grosz, *Becoming Undone: Darwinian Reflections on Life, Politics, and Art* (Durham & London: Duke University Press, 2011).
22. Alberto Pérez Gómez, *Architecture and the Crisis of Modern Science* (Cambridge, MA: MIT Press, 1983).

23. Robin Evans, "Translations from Drawing to Building and Other Essays" (1997), quoted in, Peg Rawes, "Spinoza's Architectural Passages and Geometric Comportments," in *Spinoza Beyond Philosophy*, ed. Beth Lord (Edinburgh: University of Edinburgh Press, 2012).
24. See, Peg Rawes, *Space, Geometry and Aesthetics: Through Kant and Towards Deleuze* (Basingstoke: Palgrave Macmillan, 2008).
25. For example, see: (trans. K. Montin) Luce Irigaray, *Thinking the Difference: For a Peaceful Revolution* (London: Athlone, 1994); Donna Haraway, *Simians, Cyborgs and Women: The Reinvention of Nature* (London: Free Association Books, 1991); Rosi Braidotti, *Transpositions: On Nomadic Ethics* (Cambridge: Polity Press, 2006); and, Grosz, *Becoming Undone: Darwinian Reflections on Life, Politics, and Art*.
26. Carpo, *The Alphabet and the Algorithm*, p. 130.
27. Spuybroek, *The Sympathy of Things*, pp. 293–94.
28. Beatriz Preciado, "Architecture as a Practice of Biopolitical Disobedience," in *LOG 25* (New York: Anyone Corporation, summer 2012).
29. (D. Rothenberg, trans. & ed.) Arne Naess, *Ecology, Community and Lifestyle* (Gateshead: Athenæum Press Ltd., 1995), pp. 253–54.
30. Fritof Capra writes: "Shallow ecology is anthropocentric. It views humans as above or outside of nature, as the source of all value and ascribes only instrumental or use value to nature. Deep ecology does not separate humans from the natural environment, nor does it separate anything else from it. It does not see the world as a collection of isolated objects but rather as a network of phenomena that are fundamentally interconnected and interdependent." In George Sessions (ed.), *Deep Ecology for the Twenty-first Century* (Boston & London: Shambhala Publications Inc., 1995), p. 20.
31. See: Braidotti, *Transpositions: On Nomadic Ethics*; and, Moira Gatens & Genevieve Lloyd, *Collective Imaginings: Spinoza, Past and Present* (New York: Routledge, 1999).
32. See Steven Loo, *Interstices Under Construction Symposium: Technics, Memory and the Architecture of History* (Launceston: University of Tasmania, 2011).
33. See: Haraway, *Simians, Cyborgs and Women*.
34. See: Braidotti, *Transpositions: On Nomadic Ethics*.
35. See: Grosz, *Becoming Undone*.
36. See: Lorraine Code, *Ecological Thinking: The Politics of Epistemic Location* (Oxford: Oxford University Press, 2006).
37. See: ed. Peg Rawes, *Relational Architectural Ecologies: Architecture, Nature and Subjectivity*; which includes, philosophers, Rosi Braidotti, Elizabeth Grosz and Lorraine Code, architects, Doina Petrescu and Katie Lloyd Thomas, medic Anita Berlin, and myself proposing relational socio-economic, cultural and sexed concepts of ecology.
38. See Irigaray, *Thinking the Difference*.
39. This is a three-year project funded by the UK's Arts and Humanities Research Council. Publications from the project, including podcasts by architects Alex Ely from Mae Architects, Phil Hamilton from Peter Barber Architects and

Sarah Wigglesworth are available at http://www.equalitiesofwellbeing.co.uk/publications-from-equalities-of-wellbeing-housing-workshop/.

40. See The Press Association, "Bank of England's Mark Carney Highlights Housing Market's Risk to UK," *The Guardian*, May 18, 2014.
41. See, for example, A. Clarke, S. Morris, C. Udagawa, & P. Williams, *The Role of Housing Organisations in Reducing Poverty* (The Joseph Rowntree Foundation and University of Cambridge, February 2014); and, Shelter and Crisis, *A Roof over My Head. Sustain: A Longitudinal Study of Housing Wellbeing in the Private Rented Sector*. Final Report, 2014.
42. For an explanation about how current usage of the term "affordable" in UK housing markets in fact excludes many low-income buyers or tenants, see Colin Wiles, "Affordable Housing Does Not Mean What You Think It Means," *The Guardian*, February 3, 2014, http://www.theguardian.com/housing-network/2014/feb/03/affordable-housing-meaning-rent-social-housing [accessed September 7, 2014].
43. Hilary Osborne, "London Renters Trapped in £1,000 a Month 'Rabbit Hutch Properties,'" *The Guardian*, August 29, 2014: http://www.theguardian.com/money/2014/aug/29/london-renters-trapped-1000-month-rabbit-hutch-studio-flat [accessed September 7, 2014].
44. Mayor of London's Office, London's Housing Standards Evidence Summary, *Summary of Evidence on Proposed Housing Design Standards for the Examination in Public of the Draft Replacement London Plan* (London: Mayor of London's Office, 2010).
45. See Alan Evans & Oliver MarcHartwich O.M., *Unaffordable Housing: Fables and Myths* (London: Policy Exchange and Localis, 2005).
46. Richard G. Wilkinson & Kate Pickett, *The Sprit Level: Why More Equal Societies Almost Always Do Better* (London: Allen Lane, 2009).
47. Also see Maddy Power, "*Vicious Circle: How Inequality and Our Broken Property Market Reinforce Each Other*," July 8, 2014. Available at http://www.equalitiesofwellbeing.co.uk/publications-from-equalities-of-wellbeing-housing-workshop/ [accessed September 10, 2014].
48. See, for example, the RIBA (Royal Institute of British Architects) Report, *The Case for Space: The Size of England's New Homes* (London: RIBA, 2011); and, *The Way We Live Now: What People Need and Expect from Their Homes*," an Ipsos MORI report for the RIBA (London: RIBA, 2012).
49. See "Housing Standards Review" (London: Department for Communities and Local Government, UK Government, 2013).
50. "London's Housing Standards Evidence Summary," *Summary of Evidence on Proposed Housing Design Standards for the Examination in Public of the Draft Replacement London Plan*.
51. See, for example, Andrew Drury, "Parker Morris: Holy Grail or Wholly Misguided?" in *Town and Country Planning Association Journal*, August 2008 (Ilkley: HATC, 2008).

Bibliography

Adorno, Theodor, & Horkeimer, Max, *Dialektik der Aufklärung* (Amsterdam: Querido Verlag N. V. 1947): (Jephcott, Edmund trans.), "The Culture Industry: Enlightenment as Mass Deception," in *The Dialectic of Enlightenment* (Stanford, CA: Stanford University Press, 2002)

Aggregate, *Governing by Design: Architecture, Economy, and Politics in the Twentieth Century* (Pittsburgh: University of Pittsburgh Press, 2012)

AI Lab, "Logo Memos," at http://dspace.mit.edu/handle/1721.1/5460

AIA (American Institute of Architects), "Integrated Delivery: A Guide," at http://www.aia.org/contractdocs/aias077630

Aish, Robert, & Woodbury, Robert, "Multi-level Interaction in Parametric Design," in Butz, Andreas, Fisher, Brian, Krüger, Antonio, & Olivier, Patrick (eds.), *Smart Graphics: 5th International Symposium* (Frauenwörth Cloister: Springer, 2005)

Alberti, Leon Battista (Leach, Neil, Rykwert, Joseph, & Tavernor, Robert (eds.)), *On the Art of Building in Ten Books* (Cambridge, MA: MIT Press, 1988)

Alexander, Christopher, *Notes on the Synthesis of Form* (Cambridge: Harvard University Press, 1964)

Anderson, Benedict, *Imagined Communities: Reflections on the Origin and Spread of Nationalism* (London: Verso, 1991)

Antonelli, Paola, *Design and the Elastic Mind* (New York: Museum of Modern Art, 2007)

Appadurai, Arjun, *Modernity at Large: Cultural Dimensions of Globalization* (Minneapolis: University of Minnesota Press, 1996)

Appadurai, Arjun, *The Future as Cultural Fact: Essays on the Global Condition* (London & New York: Verso Books, 2013)

"Architecture Lobby Manifesto," Venice Biennale 2014, at http://architecture-lobby.org

Arendt, Hannah, *Between Past and Future; Eight Exercises in Political Thought* (New York: Viking Press, 1968)

Arendt, Hannah, *The Human Condition* (Chicago: University of Chicago Press, 1998)

Arendt, Hannah, *The Life of the Mind* (New York: Harcourt Brace Jovanovich, 1978)

Arendt, Hannah, *The Origins of Totalitarianism* (New York: Harcourt, Brace & World, 1966)

Arrighi, Giovanni, *Adam Smith in Beijing: Lineages of the Twenty-First Century* (London: Verso, 2007)

Avanessian, Armen, & Mackay, Robin, *#Accelerate: The Accelerationist Reader* (Falmouth: Urbanomic, 2014)

Babbage, Charles (1832), "On the Economy of Machinery and Manufactures," at http://www.gutenberg.org/cache/epub/4238/pg4238.html
Badiou, Alain, *Metapolitics* (London & New York: Verso, 2005)
Banham, Reyner, *Theory and Design in the First Machine Age* (New York: Praeger, 1967)
Barkan, Joshua, *Corporate Sovereignty: Law and Governance under Capitalism* (Minneapolis: University of Minnesota Press, 2013)
Baxi, Kadambari, & Martin, Reinhold, *Multi-National City: Architectural Itineraries* (Barcelona: Actar Press, 2007)
Berardi, Franco "Bifo", *The Soul at Work: From Alienation to Autonomy* (Los Angeles, CA: Semiotext(e)/Foreign Agents, 2009)
Bergson, Henri, *Creative Evolution* (Mineola, NY: Dover Publications, 1998)
Bergson, Henri, *Time and Free Will: An Essay on the Immediate Data of Consciousness* (Mineola, NY: Dover Publications, 2001)
Bernstein, Phillip G., & Deamer, Peggy, *BIM In Academia* (New Haven, CT: Yale School of Architecture Press, 2011)
Bernstein, Phillip G., & Deamer, Peggy (eds.), *Building (in) the Future: Recasting Labor in Architecture* (New Haven: Yale School of Architecture, 2010)
Braidotti, Rosi, *Transpositions: On Nomadic Ethics* (Cambridge: Polity Press, 2006)
Brand, Stewart, *The Media Lab: Inventing the Future at MIT* (New York: Viking, 1987)
Brassier, Ray, *Nihil Unbound* (Basingstoke: Palgrave Macmillan, 2010)
Brassier, Ray, "Wandering Abstraction," in *Meta-Mute* (London: Mute, February 2014), at http://www.metamute.org/editorial/articles/wandering-abstraction
Bratton, Benjamin H., *The Stack: On Software and Sovereignty* (Cambridge, MA: MIT Press, 2015)
Brauer, Fae, & Callen, Anthea (eds.), *Art, Sex, and Eugenics* (London: Ashgate Press, 2008)
Broadbent, Geoffrey, & Ward, Anthony, "Design Methods in Architecture," *Architectural Association Paper* (New York: G. Wittenborn, 1969)
Brown, Wendy, *Walled States, Waning Sovereignty* (New York: ZoneBooks, 2010)
Bucci, Federico, & Mulazzani, Marco, *Luigi Moretti: Works and Writings* (New York: Princeton Architectural Press, 2000)
Burke, Edmund (Boulton, James T. ed.), *A Philosophical Enquiry into the Origin of Our Ideas of the Sublime and Beautiful* (Notre Dame, IN: University of Notre Dame Press, 1958)
Burry, Mark, *Scripting Cultures* (Chichester: Wiley Press, 2011)
Butler, Judith, *Bodies That Matter* (New York: Routledge, 1993)
Carpo, Mario, *The Alphabet and the Algorithm* (Cambridge, MA: MIT Press, 2011)
Castoriadis, Cornelius (Rockhill, Gabriel ed.), *Postscript on Insignificance: Dialogues with Cornelius Castoriadis* (New York: Continuum, 2001)
Castoriadis, Cornelius, "The Logic of Magmas and the Question of Autonomy," in Curtis, David Ames (ed.), *The Castoriadis Reader* (Oxford: Blackwell Books, 1997)
Chakrabarty, Dipesh, *Provincializing Europe: Postcolonial Thought and Historical Difference* (Princeton: Princeton University Press, 2000)

Cheng, Maria, "No Cows Died to Make This Burger," *Jacksonville Florida Times-Union*, August 5, 2013, at http://jacksonville.com/news/2013-08-05/story/no-cows-died-make-burger

Clarke, A., Morris, S., Udagawa, C., & Williams, P., *The Role of Housing Organisations in Reducing Poverty*, The Joseph Rowntree Foundation and University of Cambridge, February 2014

Clough, Patricia, & Willse, C. (eds.), *Beyond Biopolitics: Essays on the Governance of Life and Death* (Durham: Duke University Press, 2011)

Code, Lorraine, *Ecological Thinking: The Politics of Epistemic Location* (Oxford: Oxford University Press, 2006)

Cogdell, Christina, *Eugenic Design: Streamlining America in the 1930s* (Philadelphia, PA: University of Pennsylvania Press, 2004 & 2010)

Cogdell, Christina, "Tearing Down the Grid," *Design and Culture* 3:1 (2011)

Cogdell, Christina, *Visual Culture and Evolution* (Baltimore, MD: The Centre for Art, Design & Visual Culture UMBC, 2012)

Cogdell, Christina, & Currell, Susan (eds.), *Popular Eugenics: National Efficiency and American Mass Culture in the 1930s* (Athens, OH: Ohio University Press, 2006)

Construction Industry Task Force, "Rethinking Construction: Report of the Construction Industry Task Force," at http://www.constructingexcellence.org.uk/pdf/rethinking%20construction/rethinking_construction_report.pdf

Contemporary Architecture Practice, http://www.c-a-p.net/images_res_tower.html

Cooke, Catherine, *Russian Avant-Garde: Theories of Art, Architecture and the City* (London: Academy Editions, 1995)

Correll, Barbara, Kracauer, Siegfried, & Zipes, Jack, "The Mass Ornament," *New German Critique, No. 5* (Spring, 1975) (Durham: Duke University Press, 1975)

Cruz, Marcos, & Spiller, Neil (eds.), *Neoplasmatic Design, AD* issue 78:6 (November/December 2008)

Daston, Lorraine, & Galison, Peter, *Objectivity* (New York: Zone Books, 2007)

de Sola Pool, "Ithiel," in *Technologies of Freedom* (Cambridge, MA: Belknap Press, 1983)

Deamer, Peggy (ed.), *Architecture and Capitalism: 1845 to the Present* (London/New York: Routledge, 2013)

Deamer, Peggy (ed.), *Re-Reading Perspecta: The First Fifty Years of the Yale Architecture Journal* (Cambridge, MA: MIT Press, 2005)

Deamer, Peggy (ed.), *The Architect as Worker: Immaterial Labor, the Creative Class, and the Politics of Design* (London: Bloomsbury Academic, 2015)

Deamer, Peggy (ed.), *The Millennium House* (New York: Monacelli Press, 2004)

Deleuze, Gilles, "Postscript on the Societies of Control," in *October*, Vol. 59, winter, 1992 (Cambridge, MA: MIT Press, 1992)

Deleuze, Giles, & Guattari, Félix (Massumi, Brian trans.), *A Thousand Plateaus: Capitalism and Schizophrenia* (New York: Continuum, 1996)

Drucker, Peter F., *Post-Capitalist Society* (New York: HarperCollins Publishers, 1993)

Drucker, Peter F., *The Effective Executive: The Definitive Guide to Getting the Right Things Done* (New York: HarperCollins Publishers, 1967)

Drury, Andrew, "Parker Morris: Holy Grail or Wholly Misguided?" in *Town and Country Planning Association Journal*, August 2008 (Ilkley: HATC, 2008)

Easterling, Keller, *Extrastatecraft: The Power of Infrastructure Space* (London & Brooklyn, NY: Verso, 2014)

Eastman, Chuck, Teicholz, Paul, Sacks, Rafael, & Liston, Kathleen, *BIM Handbook: A Guide to Building Information Modeling for Owners, Managers, Designers, Engineers, and Contractors* (Hoboken, NJ: John Wiley & Sons, 2011, second edition)

Edwards, Paul N., *A Vast Machine: Computer Models, Climate Data, and the Politics of Global Warming* (Cambridge, MA: MIT Press, 2013)

Edwards, Paul N., *The Closed World: Computers and the Politics of Discourse in Cold War America* (Cambridge, MA: MIT Press, 1997)

Elwood, Claude, Shannon, Claude, & Weaver, Warren, *The Mathematical Theory of Communication* (Urbana: University of Illinois Press, 1949)

Engels, Friedrich, "The Part Played by Labour in the Transition from Ape to Man" (May/June 1876, first published in *Die Neue Zeit* in 1895)

Esposito, Roberto, *Bíos: Biopolitics and Philosophy* (Minneapolis: University of Minnesota Press, 2008)

Estévez, Alberto (Tienda, Dulce trans.), *Genetic Architectures* (Barcelona and Santa Fe, NM: SITES Books and Lumen, Inc., 2003)

Estévez, Alberto, *Genetic Architectures II* (Barcelona and Santa Fe, NM: SITES Books and Lumen, Inc., 2005)

Evans, Alan, & Hartwich, Oliver Marc, *Unaffordable Housing: Fables and Myths* (London: Policy Exchange and Localis, 2005)

Evans, Robin, "Translations from Drawing to Building and Other Essays (1997)," in Lord, Beth (ed.), *Spinoza Beyond Philosophy* (Edinburgh: University of Edinburgh Press, 2012)

Ferris, Robert, "How to Make a Hamburger Without Killing the Cow," *Business Insider*, August 6, 2013, at http://www.businessinsider.com/how-cultured-beef-is-made-2013-8

Fish, Stanley, *There's No Such Thing as Free Speech—And It's a Good Thing Too* (Oxford: Oxford University Press, 1994)

Foreign Office Architects, *Phylogenesis: FOA's Ark* (Barcelona: Actar, 2004)

Foucault, Michel (Sheridan, Alan trans.), *Discipline and Punish* (London: Penguin, 1979)

Foucault, Michel, (Sheridan, Alan trans.), *Security, Territory, Population: Lectures at the Collège De France, 1977–78* (New York: Picador, 2007)

Foucault, Michel, *"Society Must Be Defended": Lectures at the Collège De France, 1975–76* (New York: Picador, 2003)

Foucault, Michel, *The Birth of Biopolitics: Lectures at the Collège De France, 1978–79* (Basingstoke: Palgrave Macmillan, 2008)

Frazer, John, *An Evolutionary Architecture* (London: Architectural Association, 1995)

Freire, Paolo, & Papert, Seymour, "The Future of School," at http://www.papert.org/articles/freire/freirePart1.html

Galloway, Alexander R., *The Interface Effect* (New York: Polity Press, 2012)

Gatens, Moira, & Lloyd, Genevieve, *Collective Imaginings: Spinoza, Past and Present* (New York: Routledge, 1999)

Gibson, Eleanor, & Pick, Anne, *An Ecological Approach to Perceptual Learning and Development* (New York: Oxford University Press, 2002)

Gibson, James, *The Ecological Approach to Visual Perception* (Hove: Psychology Press, 1979)

Gibson, James, "The Theory of Affordances," in Shaw, Robert, & Bransford, John (eds.), *Perceiving, Acting, and Knowing* (London: Wiley Press, 1977)

Giedion, Sigfried, *Mechanization Takes Command: A Contribution to Anonymous History* (New York: Norton, 1975)

Giedion, Sigfried, *Space, Time and Architecture: The Growth of a New Tradition* (Cambridge: Harvard University Press, 1967)

Gips, James, & Stiny, George, *Algorithmic Aesthetics: Computer Models for Criticism and Design in the Arts* (Berkeley: University of California Press, 1978)

Goldman, Robert, & Papson, Stephen, *Nike Culture: The Sign of the Swoosh* (London: Sage Publications, 1998)

Grosz, Elizabeth, *Becoming Undone: Darwinian Reflections on Life, Politics, and Art* (Durham and London: Duke University Press, 2011)

Guattari, Félix (Sheed, Rosemary trans.), *Molecular Revolution: Psychiatry and Politics* (Harmondsworth: Penguin, 1984)

Guattari, Félix, (1989) (Pindar, I. & Sutton, P. trans.), *The Three Ecologies* (London: Athlone Press, 2000)

Hadid, Zaha, & Schumacher, Patrik, *Latent Utopias* (Wien: Springer, 2002)

Halpern, Orit, *Beautiful Data: A History of Vision and Reason since 1945* (Durham: Duke University Press, 2014)

Hamilton Grant, Iain, *Philosophies of Nature after Schelling* (London: Bloomsbury Academic, Transversals Series, 2008)

Hanlon, Michael, "Fake Meat: Is Science Fiction on the Verge of Becoming Fact?" *The Guardian*, June 22, 2012, at http://www.theguardian.com/science/2012/jun/22/fake-meat-scientific-breakthroughs-research

Haraway, Donna, *Simians, Cyborgs and Women: The Reinvention of Nature* (London: Free Association Books, 1991)

Hardin, Peter, & Lombardo, Paul, "North Carolina's Bold Model for Eugenics Compensation," *Richmond Times-Dispatch*, August 11, 2013, at http://www.timesdispatch.com/opinion/their-opinion/north-carolina-s-bold-model-for-eugenics-compensation/article_10ed1912-b0ea-59b0-97d3-d69d6fff7203.html

Hardt, Michael, & Negri, Antonio, *Commonwealth* (Cambridge, MA: Belknap Press of Harvard University Press, 2009)

Hardt, Michael, & Negri, Antonio, *Empire* (Cambridge, MA: Harvard University Press, 2000)

Hardt, Michael, & Negri, Antonio, *Multitude: War and Democracy in the Age of Empire* (New York: The Penguin Press, 2004)

Harman, Graham, *The Quadruple Object* (London: Zero Books, 2011)

Harvey, David, *A Brief History of Neoliberalism* (Oxford: Oxford University Press, 2005)

Harwood, John, *The Interface: IBM and the Transformation of Corporate Design, 1945-1976* (Minnesota: University of Minnesota Press, 2011)

Hayles, N. Catherine, *My Mother Was a Computer: Digital Subjects and Literary Texts* (Chicago: University of Chicago Press, 2005)

Heidegger, Martin, *The Question Concerning Technology, and Other Essays* (New York: Harper & Row, 1977)

Hensel, Michael, & Menges, Achim (eds.), *Morpho-Ecologies* (London: AA Publications, 2006)

Holweg, Mattias, "The Genealogy of Lean Production," in *Journal of Operations Management*, Vol. 25, Issue 2, March 2007 (Philadelphia, PA: Elsevier Press, 2007)

Irigaray, Luce, (Montin, K. trans.), *Thinking the Difference: For a Peaceful Revolution* (London: Athlone, 1994)

Iurascu, Ilinca, Parikka, Jussi, & Winthrop-Young, Geoffrey (eds.), *Theory, Culture, & Society, Special Issue: Cultural Techniques*, No. 30, November 2013

Jameson, Fredric, *Archaeologies of the Future: The Desire Called Utopia and Other Science Fictions* (New York: Verso, 2005)

Jameson, Fredric, "Architecture and the Critique of Ideology," in Ockman, Joan (ed.), *Architecture, Criticism, Ideology* (Princeton: Princeton Architectural Press, 1985)

Jameson, Fredric, "Is Space Political?," in Leach, Neil (ed.) *Rethinking Architecture* (London: Routledge, 1997)

Jameson, Fredric, "The Constraints of Postmodernism," in Leach, Neil (ed.) *Rethinking Architecture* (London: Routledge, 1997)

Jencks, Charles, & Valentine, Maggie, "The Architecture of Democracy: The Hidden Tradition," in *Architectural Design*, Profile 69 (London: Academy Editions, 1987)

Johnson, Corey, "Female Inmates Sterilized in California Prisons Without Approval," Center for Investigative Reporting, July 7, 2013, at http://cironline.org/reports/female-inmates-sterilized-california-prisons-without-approval-4917

Kensek, Karen, & Noble, Douglas (eds.), *Building Information Modeling: BIM in Current and Future Practice* (Hoboken, NJ: John Wiley & Sons, 2014)

Kittler, Friedrich, *Discourse Networks 1800/1900* (Stanford, CA: Stanford University Press, 1990)

Kittler, Friedrich, *Gramophone, Film, Typewriter* (Stanford, CA: Stanford University Press, 1999)

Klossowski, Pierre, *Nietzsche and the Vicious Circle* (Chicago: University of Chicago Press, 1997)

Kolarevic, Branko, *Architecture in the Digital Age: Design and Manufacturing* (New York: Spon Press, 2003)

Kracauer, Siegfried, *Das Ornament der Masse* (Frankfurt am Main: Suhrkamp, 1963)

Kracauer, Siegfried, "The Mass Ornament," *Frankfurter Zeitung*, July 9 & 10, 1927

Kuhn, Thomas, *The Structure of Scientific Revolutions* (Chicago: University of Chicago Press, 1962)

Kurgan, Laura, *Close Up at a Distance: Mapping, Technology, and Politics* (Brooklyn, NY: Zone Books, 2013)

Kwinter, Sanford, "Ecological Thinking," at *Proto/e/co/logics: Speculative Materialism in Architecture* seminar, Croatia, 2011, http://vimeo.com/28810672

Lacan, Jacques, *Écrits: A Selection* (London: Tavistock, 1977 [1959])

Laclau, Ernesto, & Mouffe, Chantal, *Hegemony and Socialist Strategy: Towards a Radical Democratic Politics* (London: Verso, 1985)

Latour, Bruno, *Reassembling the Social: An Introduction to Actor-Network-Theory* (Oxford & New York: Oxford University Press, 2007)

Latour, Bruno, *We Have Never Been Modern* (Cambridge, MA: Harvard University Press, 1993)

Latour, Bruno, & Woolgar, Steve, *Laboratory Life: The Construction of Scientific Facts* (Princeton: Princeton University Press, 1986)

Laugier, Marc-Antoine, *An Essay on Architecture* (Los Angeles: Hennessey & Ingalls, 1977 [1755])

Lazzarato, Maurizio, "Art and Work," in *Parachute 122 "Travail ** Work"*, June 2006, at http://www.parachute.ca/public/+100/122.htm

Lazzarato, Maurizio, *Immaterial Labour: Mass Intellectuality, New Constitution, Post-Fordism, and All That* (London: Red Notes, 1994)

Lazzarato, Maurizio, *Signs and Machines: Capitalism and the Production of Subjectivity* (Los Angeles, CA: Semiotext(e), 2014)

Le Corbusier (Etchells, Frederick trans.), *Towards a New Architecture* (London: Butterworth Architecture, 1989)

Leach, Neil, *Designing for a Digital World, Digital Tectonics* (London: Academy Press, 2004)

Leach, Neil (ed.), *Digital Cities AD* (Oxford: Wiley Press, 2009)

Leach, Neil, *Fabricating the Future and Camouflage* (Shanghai: Tongji University Press, 2012)

Leach, Neil, *Machinic Processes* (Beijing: China Architecture and Building Press, 2010)

Leach, Neil, "Parametrics Explained," in Oosterhuis, Kas (ed.), *Next Generation Building* (Delft: TU Delft, 2014)

Leach, Neil (ed.), *Rethinking Architecture: A Reader in Cultural Theory* (London: Routledge, 1997)

Leach, Neil, *Scripting the Future* (Shanghai: Tongji University Press, 2012)

Leach, Neil, *Swarm Intelligence* (LATP, forthcoming)

Leach, Neil, *The Anaesthetics of Architecture* (Cambridge, MA: MIT Press, 1999)

Leach, Neil, & Xu, Weiguo (eds.), *Design Intelligence: Advanced Computational Techniques for Architecture* (Beijing: China Architecture and Building Press, 2010)

Lefort, Claude, *Democracy and Political Theory* (Minneapolis: University of Minnesota Press, 1988)

Lloyd Wright, Frank, *An Organic Architecture: The Architecture of Democracy* (London: Lund Humphries, 1939)

Lloyd Wright, Frank, *When Democracy Builds* (Chicago: University of Chicago Press, 1945)

Loo, Steven, *Interstices* Under Construction Symposium: *Technics, Memory and the Architecture of History* (Launceston: University of Tasmania, 2011)

Lynn, Greg, *Animate Form* (New York: Princeton Architectural Press, 1999)

Lyotard, Jean-François, *La condition postmoderne: rapport sur le savoir* [*The Postmodern Condition: A Report on Knowledge*] (Paris: Minuit, 1979), first published in English: Lyotard, Jean-François (Bennington, Geoff, & Massumi, Brian trans.), *The Postmodern Condition: A Report on Knowledge* (Manchester: Manchester University Press, 1984)

Lyotard, Jean-Francois, *The Postmodern Condition*, Geoffrey Bennington and Brian Massumi (eds.). (Minneapolis: Minnesota University Press, 1988)

MacKenzie, Donald, *An Engine, Not a Camera: How Financial Models Shape Markets* (Cambridge, MA: MIT Press, 2006)

MacKenzie, Donald, Muniesa, Fabian, & Siu, Lucia (eds.), *Do Economists Make Markets? On the Performativity of Economics* (Princeton: Princeton University Press, 2007)

Mae Architects, Phil Hamilton from Peter Barber Architects and Sarah Wigglesworth are available at http://www.equalitiesofwellbeing.co.uk/publications-from-equalities-of-wellbeing-housing-workshop/

Manovich, Lev, *Software Takes Command: Extending the Language of New Media* (New York & London: Bloomsbury, 2013)

Marazzi, Christian, *The Violence of Financial Capitalism* (Los Angeles, CA: Semiotext(e), 2010)

Martin, Reinhold, *The Organizational Complex: Architecture, Media, and Corporate Space* (Cambridge, MA: MIT Press, 2003)

Martin, Reinhold, *Utopia's Ghost: Architecture and Postmodernism, Again* (Minneapolis: University of Minnesota Press, 2010)

Marx, Karl, *Capital, Vol.1* (Moscow: Progress Publishers, 1887)

Marx, Karl, *Critique of the Gotha Program* (written in April 1875 and first published in an abridged version in Die Neue Zeit, Volume 1, No. 18, in 1890)

Marx, Karl, *Das Kapital, Kritik der Politischen Ökonomie (Capital: Critique of Political Economy)* Volume I (Verlag von Otto Meisner, 1867)

Marx, Karl, *Economic and Philosophical Manuscripts*, 1844; found in T. B. Bottomore, *Karl Marx Early Writings* (London: C.A. Watts and Co. Ltd., 1963)

Marx, Karl (1858), *Grundrisse* (Moscow: Verlag für Fremdsprachige Literatur, 1939)

Maturana, Humberto R., & Varela, Francisco J., *Autopoiesis and Cognition: The Realization of the Living* (Dordrecht: D. Reidel Pub. Co., 1980)

Mayor of London's Office, "London's Housing Standards Evidence Summary," in *Summary of Evidence on Proposed Housing Design Standards for the Examination in Public of the Draft Replacement London Plan* (London: Mayor of London's Office, 2010)

Mbembe, Achille, *On the Postcolony* (Berkeley: University of California Press, 2001)
McGraw-Hill Construction Analytics, *Smart Market Report: The Business Value of BIM* (New York: McGraw-Hill Publications, 2009)
McLuhan, Marshal, *Understanding Media: The Extensions of Man* (New York: McGraw-Hill, 1964)
Meikle, Jeffrey, *Twentieth-Century Limited: Industrial Design in America, 1925–1939* (Philadelphia: Temple University Press, 2001)
Meillassoux, Quentin, *After Finitude* (London: Bloomsbury Academic, 2010)
Mihata, Kevin, "The Persistence of Emergence," in Eve, Raymond A., Horsfall, Sara, & Lee, Mary E. (eds.), *Chaos, Complexity, and Sociology: Myths, Models, and Theories* (Thousand Oaks, CA: Sage Publications, 1997)
Mitchell, Timothy, *Carbon Democracy: Political Power in the Age of Oil* (London: Verso, 2011)
Mitchell, Timothy, *Rule of Experts: Egypt, Techno-politics, Modernity* (Berkeley: University of California Press, 2002)
Moretti, Luigi, "Ricerca Matematica in Architettura e Urbanisticâ," *Moebius* IV:1, pp. 30–53 (1971)
Morton, Timothy, *Hyperobjects: Philosophy and Ecology After the End of the World* (Minneapolis: University of Minnesota Press, 2013)
Mostafavi, M., & Doherty, G. (eds.), *Ecological Urbanism* (Baden: Lars Muller Publishers; Harvard, MA: Harvard University Graduate School of Design, 2010)
Mouffe, Chantal, *The Return of the Political* (London: Verso, 2006)
Mumford, Lewis, *Technics and Civilization* (New York: Harcourt, Brace and Co, 1934)
Naess, Arne (Rothenberg, D trans. & ed.), *Ecology, Community and Lifestyle* (Gateshead, Tyne and Wear: Athenæum Press Ltd., 1995)
Negroponte, Nicholas, "A 30-year History of the Future," *TED talk* available online at https://www.ted.com/talks/nicholas_negroponte_a_30_year_history_of_the_future?language=en
Negroponte, Nicholas, *Soft Architecture Machines* (Cambridge, MA: MIT Press, 1975)
Negroponte, Nicholas, *The Architecture Machine: Toward a More Human Environment* (Cambridge, MA: MIT Press, 1970)
Negroponte, Nicholas, "The Origins of the Media Lab," in Wiesner, Jerome B., & Rosenblith, Walter A., *Jerry Wiesner: Scientist, Statesman, Humanist: Memories and Memoirs* (Cambridge, MA: MIT Press, 2004)
Negroponte, Nicholas, "Toward a Theory of Architecture Machines," *Journal of Architectural Education* (1947–1974) 23.2, pp. 9–12 (March 1969)
Nelson, Cary, & Grossberg, Lawrence (eds.), *Marxism and the Interpretation of Culture* (Urbana: University of Illinois Press, 1988)
Newman, Bernard, *Behind the Berlin Wall* (London: Robert Hale, 1964)
Nitzan, Jonathan, & Bichler, Shimshon, *Capital as Power: A Study of Order and Creorder* (Milton Park, Abingdon, Oxon: Routledge, 2009)
Ong, Aihwa, *Neoliberalism as Exception: Mutations in Citizenship and Sovereignty* (Durham: Duke University Press, 2006)

Oosterhuis, Kas (ed.), *Next Generation Building* (Delft: TU Delft, 2014)
Osborne, Hilary, "London Renters Trapped in £1,000 a Month 'Rabbit Hutch Properties,'" *The Guardian*, August 29, 2014: http://www.theguardian.com/money/2014/aug/29/london-renters-trapped-1000-month-rabbit-hutch-studio-flat
Papert, Seymour, *Mindstorms: Children, Computers, and Powerful Ideas* (New York: Basic Books, 1980)
Papert, Seymour, "Situating Constructionism," in Papert, S., & Harel, I. (eds.), *Constructionism* (Norwood, NJ: Ablex Publishing Corporation, 1991), at http://www.papert.org/articles/SituatingConstructionism.html
Papert, Seymour, "Teaching Children Thinking. Artificial Intelligence Memo Number 247" (MIT Cambridge, Artificial Intelligence Lab, 1971), at http://stager.org/articles/teachingchildren.html
Papert, Seymour, "Teaching Children to Be Mathematicians vs. Teaching About Mathematics. Artificial Intelligence Memo Number 249" (MIT Cambridge, Artificial Intelligence Lab, 1971), at http://hdl.handle.net/1721.1/5837
Papert, Seymour, "Uses of Technology to Enhance Education. Artificial Intelligence Memo Number 298" (MIT Cambridge, Artificial Intelligence Lab, 1973), at http://hdl.handle.net/1721.1/6213
Peck, Jamie, *Constructions of Neoliberal Reason* (Oxford: Oxford University Press, 2010)
Pérez Gómez, Alberto, *Architecture and the Crisis of Modern Science* (Cambridge, MA: MIT Press, 1983)
Petit, Emmanuel, "Involution, Ambience, and Architecture," in *Log #29 "In Pursuit of Architecture"* (New York: Anyone Corporation Press, Fall 2013)
Petro, Greg http://www.forbes.com/sites/gregpetro/2012/10/25/the-future-of-fashion-retailing-the-zara-approach-part-2-of-3/
Pickett, Kate, & Wilkinson, Richard G., *The Sprit Level: Why More Equal Societies Almost Always Do Better* (London: Allen Lane, 2009)
Picketty, Thomas, (Goldhammer, Arthur trans.), *Capital in the Twenty-First Century* (Boston: Harvard University Press, 2014)
Picon, Antoine, "Architecture and Mathematics: Between Hubris and Restraint," in Legendre, G. (ed.), *Mathematics of Space: Architectural Design*, Architectural Design Profile Number 212 (Chichester: John Wiley & Sons, 2011)
Picon, Antoine, *Digital Culture in Architecture: An Introduction for the Design Professions* (Basel: Birkhäuser, 2010)
Polanyi, Karl, *The Great Transformation* (Boston: Beacon Press, 1957)
Power, Maddy, "Vicious Circle: How Inequality and Our Broken Property Market Reinforce Each Other," July 8, 2014, at http://www.equalitiesofwellbeing.co.uk/publications-from-equalities-of-wellbeing-housing-workshop/
Preciado, Beatriz, "Architecture as a Practice of Biopolitical Disobedience," in *LOG 25* (New York: Anyone Corporation, Summer 2012)
Propublica http://www.propublica.org/article/how-we-observed-censorship-on-sina-weibo

ProPublica, "China's Memory Hole: The Images from Sina Weibo," in *China's Memory Hole: Images Erased from Sina Weibo*, November 14, 2013, at https://projects.propubica.org/weibo/

Rabinow, Paul (ed.), *The Foucault Reader* (London: Penguin, 1991)

Rancière, Jacques, "Democracy, Republic, Representation," *Constellations* 13.3, pp. 297–307 (2006)

Rancière, Jacques, *Disagreement: Politics and Philosophy* (Minneapolis: University of Minnesota Press, 1999)

Rancière, Jacques, *The Politics of Aesthetics* (Rockhill, Gabriel trans.) (London: Bloomsbury Academic, reprint edition 2013)

Rawes, Peg, *Poetic Biopolitics: Practices of Relation in Architecture and the Arts* (forthcoming 2015, co-ed.)

Rawes, Peg (ed.), *Relational Architectural Ecologies: Architecture, Nature and Subjectivity* (London: Routledge, 2013)

Rawes, Peg, *Space, Geometry and Aesthetics: Through Kant and Towards Deleuze* (Basingstoke: Palgrave Macmillan, 2008)

Resnick, Mitchel, *Turtles, Termites, and Traffic Jams: Explorations in Massively Parallel Microworlds* (Cambridge, MA: MIT Press, 1994)

Reynolds, Craig W., "Flocks, Herds, and Schools: A Distributed Behavioural Mode," *Computer Graphics* 21.4 (July 1987)

RIBA (Royal Institute of British Architects) report, *The Case for Space: The Size of England's New Homes* (London: RIBA, 2011), and, "The Way We Live Now: What people need and expect from their homes," an Ipsos MORI report for the RIBA (London: RIBA, 2012)

Robinson, Kim Stanley, *Blue Mars* (New York: Bantam Books, 1996)

Robinson, Kim Stanley, *Green Mars* (New York: Bantam Books, 1994)

Robinson, Kim Stanley, *Red Mars* (New York: Bantam Books, 1993)

Rudofsky, Bernard, *Architecture without Architects: An Introduction to Nonpedigreed Architecture* (New York: Museum of Modern Art; distributed by Doubleday, Garden City, NY, 1964)

Ruskin, John, Cook, Tyas, Edward, Sir, & Wedderburn Dundas Ogilvy, Alexander, *Biblioteca Pastorum: The Economist of Xenophon; Rock Honeycomb; The Elements of Prosody and A Knight's Faith*, Chapter IV, (London: G. Allen Publisher, 1907)

Russell, Bertrand, *Introduction to Mathematical Philosophy* (London: George Allen & Unwin, Ltd./New York: The Macmillan Co., first published 1919, second edition 1920)

Saez, Emmanuel, & Piketty, Thomas, "Top Incomes and the Great Recession: Recent Evolutions and Policy Implications," in *International Monetary Fund Economic Review #61*: Vol. 3, 2013, at http://eml.berkeley.edu/~saez/piketty-saezIMF13topincomes.pdf

Sassen, Saskia, *Expulsions: Brutality and Complexity in the Global Economy* (Cambridge, MA: The Belknap Press of Harvard University Press, 2014)

Sassen, Saskia, *The Global City: New York, London, Tokyo* (Princeton: Princeton University Press, 1991)

Scheurer, Fabian, & Stehling, Hanno, "Lost in Parameter Space?" *Architectural Design* 81 4, pp. 70–79 (2011)

Schmitt, Carl, *The Nomos of the Earth in the International Law of the Jus Publicum Europaeum* (New York: Telos Press, 2003)

Schumacher, Patrik, "Parametricism: A New Global Style for Architecture and Urban Design," *Architectural Design*, 79.4, pp. 14–23 (2009)

Schumacher, Patrik, "Parametricism: A New Global Style for Architecture and Urbanism," in Leach, Neil (ed.), *Digital Cities AD* (London: John Wiley & Sons, 2009).

Schumacher, Patrik, "Parametricism as Style—Parametricist Manifesto," at http://www.patrikschumacher.com

Schumacher, Patrik, "Patrik Schumacher on Parametricism—'Let the Style Wars Begin'", *Architects Journal* 231.16, pp. 7–48 (May 6, 2010)

Schumacher, Patrik, *The Autopoiesis of Architecture, Vol.1: A New Framework for Architecture* (London: John Wiley & Sons Ltd., 2010)

Schumacher, Patrik, *The Autopoiesis of Architecture, Vol.2: A New Agenda for Architecture* (London: John Wiley & Sons Ltd., 2012)

Schumacher, Patrik, "The Parametricist Epoch: Let the Style Wars Begin," *The Architects' Journal* 231.16, pp. 41–45 (May 2010)

Schumacher, Patrik, & Eisenman, Peter, "I Am Trying to Imagine a Radical Free-market Urbanism," in Eisenman, Peter E., & Vidler, Anthony (eds.), *Log 28 - Stocktaking* (New York: S.I. Anyone Corporation, 2013)

Scott, Felicity, "Discourse, Seek, Interact," in Dutta, Arindam (ed.), *A Second Modernism: MIT, Architecture, and the "Techno-Social" Moment* (Cambridge, MA: MIT Press, 2013)

Scott, Felicity Dale Elliston, *Architecture or Techno-utopia: Politics after Modernism* (Cambridge, MA: MIT Press, 2007)

Scully, Vincent, *Modern Architecture: The Architecture of Democracy* (New York: George Braziller, 1974)

Sessions, George (ed.), *Deep Ecology for the Twenty-first Century* (Boston & London: Shambhala Publications Inc., 1995)

Shaw, Robert, & Bransford, John (eds.), *Perceiving, Acting, and Knowing* (London: Wiley, 1977)

Shelter and Crisis, *A Roof over My Head. Sustain: A Longitudinal Study of Housing Wellbeing in the Private Rented Sector*. Final Report, 2014

Shvartzberg, Manuel, "Foucault's 'Environmental' Power: Architecture and Neoliberal Subjectivization," in Deamer, Peggy (ed.), *The Architect as Worker: Immaterial Labor, the Creative Class, and the Politics of Design* (London: Bloomsbury Academic, 2015)

Siegert, Bernhard, *Cultural Techniques: Grids, Filters, Doors, and Other Articulations of the Real* (New York: Fordham University Press, 2014)

Simmel, Georg, "The Stranger," in Wolff, Kurt H. (ed. & trans.), *The Sociology of Georg Simmel* (New York: Free Press, 1950)

Sinclair, Nathalie, *The History of the Geometry Curriculum in the United States* (Charlotte, NC: Information Age Publications, 2008)

Smith, Adam, (1776), *An Inquiry into the Nature and Causes of the Wealth of Nations*, at http://www.econlib.org/library/Smith/smWN1.html#B.I,%20Ch.1,%20Of%20the%20Division%20of%20Labor

Smith, Adam (Garnier, Germain ed.), *Recherches sur la nature et les causes de la richesse des nations; traduction nouvelle, avec des notes et observations par Germain Garnier* (Paris: Agasse, 1802), 2 volumes

Smith, Daniel W., & Somers-Hall, henry (eds.), *The Cambridge Companion to Deleuze* (Cambridge: Cambridge University Press, 2012)

Smith, Rick, *Technical Notes from Experiences and Studies in Using Parametric and BIM Architectural Software*, published March 4, 2007, at http://www.vbtllc.com/images/VBTTechnicalNotes.pdf

Solomon, Cynthia Solomon http://logothings.wikispaces.com/

Spinoza, Baruch, (1677) (Feldman, S. ed.), *Ethics: Treatise on the Emendation of the Intellect and Selected Letters* (Indianapolis: Hackett Publishing, 1992)

Spinoza, Baruch, *Tractatus Theologico-politicus; Tractatus Politicus* (London: George Routledge, 1990)

Spuybroek, Lars, *The Sympathy of Things: Ruskin and the Ecology of Design* (Rotterdam: nai010 publishers, 2012)

Stedman Jones, Daniel, *Masters of the Universe: Hayek, Friedman, and the Birth of Neoliberal Politics* (Princeton: Princeton University Press, 2012)

Stengers, Isabelle, *Cosmopolitics* (Minneapolis: University of Minnesota Press, 2010)

Sutherland, Ivan E., "An Electro-Mechanical Model of Simple Animals," in Sutherland, Ivan E. (ed.), *Computers and Automation* (Cambridge, MA: MIT Press, February 1958)

Sutherland, Ivan E., *Sketchpad: A Man-Machine Graphical Communication System* (PhD dissertation) (Cambridge, MA: Department of Electrical Engineering, MIT, 1963)

Sutherland, Ivan E., "Stability in Steering Control," *Electrical Engineering*, April 1960

Tafuri, Manfredo, & Francesco Dal Co., *Modern Architecture* (New York: Abrams, 1979)

Taylor, Frederick Winslow, *Principles of Scientific Management* (London: Forgotten Books, 1911)

Teresko, John, "Parametric Technology Corp.: Changing the Way Products Are Designed," *Industry Week*, December 20, 1993

Teyssot, Georges, *A Topology of Everyday Constellations* (Cambridge, MA: MIT Press, 2013)

The National Commission on Excellence in Education, *A Nation at Risk: The Imperative for Educational Reform: A Report to the Nation and the Secretary of Education, United States Department of Education* (Washington, DC: National Commission on Excellence in Education, 1983)

The Press Association, "Bank of England's Mark Carney Highlights Housing Market's Risk to UK," *The Guardian*, May 18, 2014

Turing, Alan M., "Computing Machinery and Intelligence," *Mind*, New Series, 59.236, pp. 433–460 (October 1950)

Turner, Fred, *From Counterculture to Cyberculture: Stewart Brand, the Whole Earth Network, and the Rise of Digital Utopianism* (Chicago: University of Chicago Press, 2006)

Upitis, Alise, "Nature Normative: The Design Methods Movement, 1944–1967" (PhD diss., Cambridge, MA: MIT, 2008)

Vehlken, Sebastian, "Zootechnologies: Swarming as a Cultural Technique," in Winthrop-Young, Geoffrey, Iurascu, Ilinca, & Parikka, Jussi (eds.), *Theory, Culture, & Society, Special Issue: Cultural Techniques* (London & New York: Sage Journals, November 2013)

Virno, Paolo, *A Grammar of the Multitude: For an Analysis of Contemporary Forms of Life* (Cambridge, MA: Semiotext(e), 2003)

Walter, William Grey, "An Imitation of Life," *Scientific American* 182.5 (May 1950) WeiboScope http://weiboscope.jmsc.hku.hk/datazip/

Wentworth Thompson, D'Arcy (1915), *On Growth and Form* (Cambridge: Cambridge University Press; New York: Macmillan Press, 1945)

West, Jessamyn, *Five Technically Legal Signs for Your Library*, December 19, 2002, at http://www.librarian.net/technicality.html

Whitehead, Alfred North, *The Concept of Nature* (Amherst, NY: Prometheus Books, 2004) "Who Builds Your Architecture?" http://whobuilds.org/ "Who's Building the Guggenheim Abu Dhabi?" http://gulflabor.org/

Wiener, Norbert, *Cybernetics; or, Control and Communication in the Animal and the Machine* (New York: MIT Press, 1961)

Wiles, Colin, "Affordable Housing Does Not Mean What You Think It Means," *The Guardian*, February 3, 2014, http://www.theguardian.com/housing-network/2014/feb/03/affordable-housing-meaning-rent-social-housing

Wisnioski, Mathew, "Centrebeam: Art of the Environment," in Dutta, Arindam (ed.), *A Second Modernism: MIT, Architecture, and the "Techno-Social" Moment* (Cambridge, MA: The MIT Press, 2013)

Woodbury, Robert, *Elements of Parametric Design* (Abingdon: Routledge Press, 2010)

Zhu, Tao, et al., "The Velocity of Censorship: High-Fidelity Detection of Microblog Post Deletions," in *Proceedings of the 22nd USENIX Conference on Security*, SEC'13 (Berkeley: USENIX Association, 2013)

Index

AA Design Research lab 26
Accelerationism 13, 87
Accelerationist Manifesto 87
Accountability 110, 192, 208
Activism 19–20, 64, 193
Activist 21–22, 91, 125, 160–161, 164–165, 199
Actor Network Theory 17, 167, 171, 177
AD Design Magazine 186
Adaptability 115
Adorno, Theodor 51
Advanced Data Visualization Project (GSAPP) 161–162
Advanced Research Projects Agency (ARPA) 101, 120
AEC (Architecture Engineering Construction) 179, 187
Aesthetic/Aesthetics 2, 5, 6, 9, 13, 17, 39, 44, 46, 54, 59, 61–62, 63–64, 66, 69–70, 71, 73–74, 75, 76, 77, 79–80, 87, 91, 96–97, 106, 107, 108, 114, 121, 125, 130, 140, 155, 163, 164, 165, 179, 180, 182, 185, 187, 190, 191, 194–195, 197, 201, 215, 217, 222, 223, 229
Affect 48, 54, 96, 111, 140, 150, 220–222
Affordability 198
Affordable housing 194, 216, 230
Affordances 58, 73, 77
Agamben, Giorgio 10, 221
Agent Based Systems 29
Agnotology 214, 227
AI.implant 31
Alberti, Leon Battista 78, 201
Alexander, Christopher 8, 17, 157
Algorithm 5, 15–16, 40, 79, 82, 86, 88, 127, 130–131, 159, 165, 178, 196–197, 203, 205–207, 210–211, 216, 218, 228–229; Eugenic 131–132; Genetic 127, 130–131; Simulation 207

Algorithmic 4–5, 16–17, 70, 80–81, 83–92, 138–139, 143, 179, 204, 214
Alienation 18, 145, 148–149, 152
Alterity 83, 223
Amazon 86–87
Anarchist 132
Anderson, Benedict 114, 121
Animation 6, 69, 118, 203
Antagonism 189, 191
Anthropocene 87
Anthropology 9
Anthropomorphism 53, 218, 222–223, 225
Anti-Enlightenment 214
Anti-essentialist 10
Antihumanism 53–55
Antipublic 190
Antonelli, Paola 134, 136–137
Apartheid 65
Appadurai, Arjun 14, 18
Application Protocol Interface (API) 160
Architectural Association, The 17, 68, 129, 136–137
Architecture Machine Group 96, 107, 109, 118
Archive 159–167
Arendt, Hannah 9, 17, 122
ARPA (Advanced Research Projects Agency) 101, 120
Art Deco 125
Art Nouveau 24
Artificial Intelligence 8, 91, 96, 102, 118–119, 133, 166
Artificial Life 100, 113, 117, 133
Assemblage 4, 83, 106, 171
Assets 193
Attributes 4, 8, 55, 97, 115, 179–181, 196, 221–222
Austin, J.L. 49
AutoCAD 46, 202–203
Autocratic 84, 191

Automation 119, 138, 140
Autonomia 13
Autonomy 8, 18, 21, 51, 90, 97, 175, 190–191, 197–198, 213, 215
Autopoiesis 8, 14–17, 23, 32–34, 43, 51, 56, 89
Autopoietic 20–22, 43, 51–52, 54, 89
Avant-garde 5, 33–34, 69, 76, 141–142, 177, 179, 182, 190

Babbage, Charles 144, 157
Badiou, Alain 13, 52–55, 57
Bare life 221
Baroque 42
Bartlett School of Architecture, The 127
Base, Market 150
Base, Capitalist 182, 188
Bataille, Georges 60, 64, 76
Bateson, Gregory 92, 220
Bauhaus, The 23, 51, 55
Beautification 191, 195, 199
Beauty 31, 223
Beaux Arts 42, 45
Beer, Stafford 84
Behavioural psychology 8, 12
Benjamin, Walter 9, 79, 92–93
Bentham, Jeremy 66–67, 76
Berardi, Franco "Bifo", 13, 18
Big data 2, 6, 16, 89, 200, 209
BIM (Building Information Modeling) 2, 16, 178–179, 181–183, 187, 203–206, 210–211, 216
Biodiversity 214, 218–220, 222–224
Biological evolution in architecture 215
Biological forms 217
Biological morphologies 214
Biological semiotics 221
Biometric management 221
Biophysical sciences 218
Biopolitical 214, 220–221, 223, 225–226, 229
Biopolitical crisis 226
Biopolitical ecological thought 221
Biopolitics 10, 17, 120, 220
Bios (philosophy) 220–221, 223
Bitcoin 90–91
Blob architecture 6
Blogs 163
Bottom-up 20, 42–43, 115, 123, 133, 191–192, 196–199, 217

Braidotti, Rosi 220–221, 223–225, 229
Brain 140, 146
Branding 180, 183–184
Brown, Wendy 232
Builder 47, 201, 203, 206, 208, 210–211
Building 5, 15–16, 24, 26, 43, 45–46, 55–56, 58–72, 75–78, 86, 92, 113, 123, 128–129, 131, 133–134, 145, 154–155, 170–171, 175, 178, 186–187, 194–195, 197–198, 201–211, 213–215, 222, 225–227, 229
Burke, Edmund 97, 114, 118, 121
Burry, Mark 6, 15–16
Business 1, 11, 33, 38, 111–113, 137, 180, 182, 211–212
Butler, Judith 10, 221, 228

CAD (Computer Aided Design) 202–203, 205
Calculability 47, 177
Calculation 5–6, 13, 45, 54, 90, 139–140, 147, 153–155, 157, 168–169, 171–173, 175–177; Collective 172–173
Calculus 215–216, 222
Cantor, Georg 48
Capital 56, 80, 89, 97, 108, 114, 117, 132, 150–153, 156–157, 168, 178, 191, 217, 228
Capitalism 1, 13, 18, 24, 85, 87, 96, 113–115, 143, 145, 151, 182–183, 186, 191, 228; Financial 18, 238; Hyper-, 191; Late 1; Post-, 184
Capra, Fritjof 196, 229
Carbon emissions 214
Carpo, Mario 15–16, 201, 211, 215, 223, 228–229
Cartesian coordinate system 105
Cartesian Geometry 102
Castoriadis, Cornelius 51–53, 56–57
Catia (computer program) 70
Catts, Oron 128, 134–135, 137
Ceaușescu, Nicolae 59–60
Cell phones 171
Censor, Human 159, 161, 163–165
Censorship 159–160, 162–166
Cerf, Vint 83
China 71, 91, 133, 159, 165–166, 194
Chinese Government 160
Citizens 11, 62, 143, 158, 160

Index 247

Citizenship 12, 15
City 35–37, 55, 62, 76, 93, 108, 121, 123, 125, 136–137, 170–173, 176, 186, 190, 192–198
City, The neoliberal 190
Civic participation 194
Class (political) 47–48, 117, 125, 196
Client 22–23, 41–42, 46, 92, 134, 186, 197, 210
Climate change 213–214, 216, 227
Climate Modeling 209
Clinton, Hillary 79–80, 83, 92
Cloud 2, 80, 82–84, 204–206
Cloud Computation 205
CNC (milling machine) 71, 77
Code (computer) 5, 16, 50, 54, 62, 129, 132, 135, 152, 154, 156, 166, 178, 206, 217–218, 220
Code, Lorraine 224, 227, 229
Coexistence 192
Cognitive Science 8, 96, 101
Cohabitation 192
Colebrook, Claire 220
Collaboration 7, 107, 115, 129, 133, 163, 166, 186, 189, 205
Collective, Social 43, 53, 55, 100, 115, 117, 152, 192–194, 198
Collective, Procedure 31–33, 171, 172–173, 177
Collective Calculation 172–173
Command 16, 50, 54–55, 60, 103, 106
Commodity 13, 47, 49, 139–140, 145, 148, 151–152, 195, 210
Common notions 221–222
Commonality 218, 222
Commons, The 222
Communism 85
Communist Manifesto 43
Community 14, 114–115, 118, 145, 170, 172, 192, 198, 213, 225, 227, 229
Complexity 20, 22, 33, 36, 38–40, 42, 69, 115, 127, 129, 135, 141, 197, 203, 217, 223
Composition 10, 26, 29–31, 39, 42, 80–81, 89, 171–172, 175, 177, 196, 207, 227
Computation 5–8, 13, 45, 80, 83, 85–87, 90, 92, 127, 129, 131–135, 137, 140, 159, 202, 204–207, 209, 216; Analysis 206; Cloud 205

Computational DNA 123, 129, 135, 137, 218
Computational scripts 219
Computationally Driven Emergence 20, 80, 89–91, 100, 115, 207, 214, 216, 218
Computer Aided Design (CAD) 137
Computer programming 96, 102, 105
Computer science 47, 85, 96, 101, 118, 131
Computer, Personal 112
Computer-controlled Fabrication 203
Conatus 219, 222
Consensus 51, 176, 189–191
Conservative 88, 132
Consumerism 47, 62, 137, 148, 150, 151, 191
Constructionism 46–47, 102, 106, 119
Constructivism 43
Consultants 201
Consumption 37, 135, 149, 156, 183–184
Contemporary Architecture Practice 123, 136
Contingency 47, 51, 53, 86
Contract 49
Corporate Lobbying 193
Corporation 11, 14–16, 38, 93, 111–113, 119, 121, 152, 158, 229
Creativity 29, 46, 51–52, 92, 97, 103, 115, 117, 140, 160, 194, 198–199
Crisis 33–34, 37, 69, 110, 190–193, 195, 214, 218, 222, 225–228, 230; Biopolitical 226; Cultural 195
Cruz, Marcos 127, 136
Crypto-political 190
Cybernetics 8, 12, 17, 84, 101, 151
Cyberwar 54, 86

Daily life 168
DARPA (Defense Advanced Research Projects Agency) 101, 108, 111–113
Darwin, Charles 215, 223–224
Darwinian 125, 129, 132, 146, 214, 228–229
Data 2, 5–6, 14, 16, 38, 89–90, 108, 133, 140, 153–155, 160–167, 172, 182, 200–204, 206, 209, 216, 226; Big 2, 6, 16, 89, 200, 209; Mining 209; Search 74, 77, 91, 159–160, 163–165; Storage 206

Data Set 160, 166; Visual 160
Database 88, 160, 163, 165–166, 200, 204
de Sola Pool, Ithiel 111, 121
De-acceleration, Cultural 87
De-politicization 192
De-signification 7
Decentralization 37, 83–84, 110, 113, 117
Decentralization, Era of 113
Deconstructivism 5, 24, 39, 41, 69
Deep Ecology 196, 220, 223, 229
DeLanda, Manuel 217
Deleuze, Gilles 13, 77, 82, 152, 155, 158, 182–183, 187–188, 220, 228–229
Deliberative process 173
Demo (presentation) 114–115
Democracy 10, 17, 21, 62, 69, 76, 143, 189, 192, 197; Liberal 189; Radical 17, 51; Representative 21; Social 10, 69, 192, 197
Democratization 19, 33, 195
Demos 114
Density 35, 44, 198
Deregulation 11, 85, 97, 110, 117, 143, 192
Derrida, Jacques 9
Desargues, Girard 215
Design Methods movement 8, 17, 50, 232, 244
Design Options 207, 209
Desire 21, 37, 51, 105–106, 108, 117, 127, 148, 150, 152, 158, 168, 181–182, 188, 194, 220
Deterritorialization 183
Developer's Pro-forma 197
Development, Urban 20, 33, 36–39, 42, 133, 193–195, 198–199, 225
Device 8, 38, 67, 105, 171–172, 174, 177, 198, 206
Dewey, John 101
Dialectical 118
Dictatorship 191
Digital Era 201
Digital; Fabrication 205; Mediation 173; Metaphors 221; Representation 204; Vector 38, 42, 98, 196, 202
Digital Project (software) 70
Digitalization 176
Disciplinary Hybridization 186
Discourse genealogy 4
Disequilibrium 218, 225

Disney Hall (Los Angeles) 72
Dissident 163, 165, 177
Dissimilarity 214, 218, 220, 222, 226
Diversification 33, 37, 221
Diversity 20, 29, 34, 36–38, 125, 213–214, 216, 221, 227
Divine power 220
Divine, The 50, 154, 219–220
Division of Labor 21–22, 87, 144–147
DNA 123, 127, 129, 132, 134, 137, 190, 217–218; Computational DNA 123, 129, 135, 137, 218
Dogma 29–30, 46, 127, 129, 142, 154
Dollens, Dennis 127, 134
Drafting Instruments 201
Drafting Software 201
Drawing 7, 45–46, 74, 80, 83, 96, 98, 105–106, 137, 201–206, 222, 229; Orthographic 201
Drucker, Peter 185, 188
Dubai 123, 194
Duchamp, Marcel 53
Duration 172, 222
Dystopia 196

EAs (Eugenic Algorithms) 131–132
Ecological disaster 196
Ecological Parametricism 215
Ecological thinking 196, 221, 223–224, 228–229
Ecology 13, 113, 196, 220–224, 228–229; Deep Ecology 196, 220, 223, 229
Ecology of mind 220
Economics 2, 4, 10–12, 47, 53, 79, 81, 84–86, 88, 90–91, 133–134, 142, 192, 195, 210
Economies of scope 20
Economy 9–10, 19–21, 23–24, 35, 37, 90, 98, 109, 112, 114–115, 117, 132, 139, 144, 146, 148, 156–158, 168, 170, 179, 181, 185–186, 188, 190, 193–194, 196–197, 209, 215; Knowledge 35, 181, 185–186; Neoliberal political 190; Political 9, 156, 190, 196–197
Ecosophy 221
Ecosystem 131, 168, 171, 176
Education 10, 21, 96, 106, 109–110, 114, 116–117, 119, 121, 152, 226
Edwards, Paul 108, 120

Efficient Markets Hypothesis 12
Ego 103, 105, 180, 182
Eisenhower, Dwight 101
Eisenman, Peter 15, 93, 204
Elevation 66, 201
Ellison, Larry 200, 211
Emergence, Computationally Driven 20, 80, 89–91, 100, 115, 207, 214, 216, 218
Empiricism 12, 108
Emulation 216
Energy 85, 88, 133, 206–207, 216
Engels, Friedrich 146, 156–157
Engineering 2, 5–6, 35, 44, 83, 100, 107, 118–119, 127, 133–135, 205–206, 211
Engineers 101, 127, 129, 134, 187, 201, 203, 206–207, 211
Enlightenment 9, 45, 214, 218–220, 222
Entrepreneur 23, 38, 42, 106, 180
Environment 7, 12, 23, 26, 31–32, 35–36, 38–40, 44, 79, 103, 108, 113, 116–117, 120, 123, 127, 129, 135, 137, 141, 155, 158, 176–177, 194–195, 198, 209, 216–218, 221, 224–225, 227, 229
Environmental 6, 8, 12, 32, 108, 125, 127, 132–133, 184, 194–195, 198, 213, 215–218, 225, 227; Collapse 8; Health 225
Epigenetics 127, 132, 218
Epistemological regime 175
Epistemology 1, 10, 13, 43, 84, 102–103, 203
Epochal Style 5, 23, 32, 34, 37, 40–41
Equality 10, 12, 226
Equality Trust, The 226
Equation 48, 54, 56, 201–203
Era, Digital 201
Era of decentralization 113
ESARQ (Universitat International de Catalunya, Barcelona) 127, 134
Essentialism 53
Essentialist 224
Estévez, Alberto 127, 134–135, 137
Ethics 10, 88, 132, 135, 214, 220, 222, 224, 227–229
Ethics of care 224
Euclidean Geometry 50
Eugenic 125, 131–132, 134–136
Eugenic Algorithms 131–132
Eugenics 125, 131–132, 136

Evans, Robin 222, 229–230
Event; Historical 8; Philosophical 52–54, 57; Scenario 30–32; Symbolic 10
Evolutionary biological morphology 218
Evolutionary design 115, 127, 136
Exchange 11, 49, 82–84, 139, 145, 159–160, 198, 209
Exclusion 110, 225
Exhilaration 216
Expression 52, 54, 59–60, 75, 80–81, 86–87, 123, 131, 135, 151, 218–222
Expressionism 24

Fabrication 6, 14, 47, 56, 71, 73, 77, 203, 205; Computer-controlled Fabrication 203; Digital Fabrication 205
Facebook 159
Facilities Management 205
Factory 51, 138, 152–153, 158
Failure 21, 44, 64, 82, 175, 182, 188, 208, 210
Fairness 12
Fantasy 46, 184
Fascism 62
FBI 165
Feedback 132–133, 138, 200, 206
Fees 172–173, 186, 210; Percentage of construction 186
Feminist 214, 217, 220–221, 223–225
Fiduciary Responsibility 186
Finance 54
Financial; Argument for parametric modeling 16; Assets 193; Capitalism, Violence of 18, 238; Crisis 34; Deregulation 11; Efficiency 154; Interest 38, 42, 111; Markets 225; Models 6, 238; Mortgages 226; Returns 217; Technocracies 2; Value 209
Finitude 54
Fitness 125, 129, 131, 215
Flexibility 7, 16, 38, 138–139, 154
Flock 94–95, 98–100, 118–119
Foreign Office Architects (FOA) 123–125, 127, 136
Forensics 194
Form 3, 13–18, 29–32, 36–37, 47–49, 55, 60, 63–69, 73–77, 80–82, 86–93, 100–101, 103, 106, 127, 129–131, 135–136, 139–140, 149, 152–156, 159, 163,

171–172, 175–176, 178–182, 188, 190, 195–198, 200–203, 205–206, 214–220, 222–223, 225–228; Inert 219
Form, Production of 175
Form-driven Scripting 178
Formal Rationalization 5
Formalism 81, 182, 197
Formalist aesthetic criteria 215
Forms, Social 6, 82, 220
Forms of life 18, 101
Formula 16, 26, 31, 54, 203, 214
Foster + Partners, Norman 63
Foucault, Michel 9–12, 17, 58, 67–68, 76–77, 89, 106, 120, 152, 158, 220
Frampton, Kenneth 64
France 17, 57, 120, 157, 170
Frankfurt School, The 51
Fraternity 12
Frazer, John 8, 17, 129–130, 133, 136–137
Free market 19, 36–37, 39, 42, 93
Freedom 6, 12, 30, 37, 39–41, 67, 111, 116–117, 120–122, 154, 216
Freire, Paolo 116–117, 122
Freud, Sigmund 103, 187
Friedman, Milton 11, 15
Functionalism 51, 115
Funding, The politics of 108

Gan, Alexei 61
Gandhi, Mahatma 87
Gatens, Moira 223, 229
Gehry & Partners 70–72
Gehry, Frank 6, 70–72, 79
Generation X 179
Generation Y 179
Generative architecture 127, 129
Generative design principles 215
Genetic 47, 102–103, 123, 125, 127, 129–134, 137, 215–216, 218, 220
Genetic Algorithms 127, 130–131
Genetic Engineering 127, 134
Gentrification 199
Geoengineering 86
Geography 84
Geometric thinking 214–215, 218–219, 222, 224
Geometry 50, 80, 90, 96, 98, 102–103, 119–120, 149, 153, 202–206, 215, 219–220, 222, 224, 229; Projective 215; Cartesian 102; Euclidian 50; Turtle 96, 98, 102–103
Geophilosophy 220
Geopolitical 79–82, 88–89, 91, 112
Geopolitics 80–81, 83, 85
Gibson, Eleanor 78
Gibson, James 77
Ginzburg, Moisei 61
Global market 42, 109
Global North 216
Global South 216
Global Warming 193
Globalization 13, 19–20, 37, 43, 88, 91, 133, 158
Gmail 163
Godard, Jean-Luc 2
Google 2, 6, 74, 77, 84–86, 91, 163, 187
Gothic Revival 45
Governance 12, 80, 92, 154, 193, 198, 220–221, 226; Institutions of 193
Government 11, 51, 63, 65, 108, 112, 120, 147, 159–160, 191, 193, 198–199, 226–227, 230; Chinese 160
Gramsci, Antonio 9
Graphical User Interface (GUI) 82
Graphics 15, 90, 95–96, 98, 102, 118–119, 204
Grasshopper (computer program) 70, 183, 205, 214
Great Depression, The 192
Great Firewall (of China), The 159, 161–162
Greed 192
Grosz, Elizabeth 224–225, 228–229
Growth, Uneven 193
Guattari, Félix 13, 151, 158, 182–183, 187–188, 220–221, 228
Guggenheim Museum (Bilbao) 72
GUI (Graphical User Interface) 82, 90

Habermas, Jürgen 9
Habitat 121, 221
Habitus 221
Haeckel, Ernst 221
Halliburton 87
Haraway, Donna 220–221, 224–225, 229
Harman, Graham 217
Happenings 53
Hayek, Friedrich 11, 15, 83

Health 10, 144, 212, 225
Health, Environmental 225
Health, Mental 144, 225
Healthcare 216, 226
Hegemony 17, 40, 69, 108, 110, 139, 168, 189
Hensel, Michael 215, 228
Herzog & de Meuron 133
Heterogeneity 176
Heuristics 29–31, 142–143, 204, 207
Higher education 117
Hipster 178–179
Historicism 53
Hobbes, Thomas 9
Holland, John 129
Hollwich, Mattias 128, 134–135
Homogenization 195
Horkheimer, Max 51
Housing 121, 123, 136, 171, 190, 194, 198, 216, 218, 225–227, 230
Housing, Affordable 194, 216, 230
Housing crisis 218, 227
Hubris 215–216, 228
Human, The 17, 55, 133, 137, 165, 217, 227
Human-nature relations 214, 217
Humanism 50, 143, 146, 151
Hybridization, Disciplinary 186
Hyper-aesthetics 195
Hyper-capitalism 191

Identity 12, 20, 37, 42, 69, 123, 175, 196, 224
Ideology 2, 12, 39, 48, 54, 62, 76, 117, 123, 125, 127, 129, 131–133, 135, 137–138, 140, 142–143, 146, 148, 151, 154, 197; Critique 117, 140, 142–143, 148, 151, 154, 182, 197
Illegality 192
Imaginary 2, 51–52, 55–56, 86, 91, 94, 96–98, 100, 115, 117, 180, 187
Immanence 7, 16
Immanent biodiversity 219; Differentiation 220
Immaterial labour 151, 158
Immediacy 164
Immigration 125, 177
Individuation 105, 115
Inequality 2, 190, 192, 225–226, 230

Inert Form 219
Inert Matter 219
Infinitude 222
Informality 115, 194, 198
Information age 109, 111, 115, 117, 119
Information theory 8
Infrastructure 2, 14, 83–84, 91, 98, 114–115, 163, 194, 198–199, 205–206, 214
Infrastructure-as-a-service 163
Inhuman, The 54
Inhumane 221
Innovation 2, 8, 10, 19–20, 35, 37, 40, 69, 71, 79, 110–111, 113, 141, 148–151, 169–170, 174, 176, 179, 183, 210, 214–216, 224
Inputs, Recursive streams of 6
Insurance 113, 210
Integrated Project Delivery 210
Intelligence Management 178
Interaction, Design 15, 29, 35–36, 83–84, 88, 90, 92, 208, 211
Interaction, Social 13, 32, 35–36, 46, 52, 88, 129, 141, 171–172, 175, 208
Interest, Private 11, 22–23, 42, 97, 103, 115, 117, 193
Interest, Public 22–23, 198
Interest, Financial 38, 42, 111
Interface 5, 12, 35–36, 73, 79–84, 86–87, 90, 103, 107, 160, 162–164, 198–199, 204
Interfacial 81–82
International Style 5, 195
Internet of Things 81
Investment 49, 84, 173, 193, 195
Invisibilization 114
Irigaray, Luce 224, 229
Iron Curtain, The 24, 43
Iteration 5, 8–9, 12, 105, 141, 207

Jameson, Fredric 59–61, 63–64, 76
Jencks, Charles 62, 76
Journalists 160
Jurisdiction 92, 198

Kant, Immanuel 217, 229
Kay, Alan 106
Keynes, John Maynard 83
Khrushchev, Nikita 83–84

Kittler, Friedrich 46
Klein, Melanie 178–179, 181–182, 184, 187–188
Knowledge Economy 35, 181, 185–186
Koolhaas, Rem 82, 197
Kuhn, Thomas 51
Kushner, Marc 128

Labor 15, 56, 89, 144–148, 180, 186, 197, 232; Cheaper 110; Division of 21–22, 87, 144–147; Fordist 145–148; Immaterial 150–152, 155, 233; Intellectual 45; Post-Fordist 13, 138, 142, 150–152, 154, 156
Labor (birthing) 131
Lacan, Jacques 181, 184, 187–188
Laclau, Ernesto 9, 17
Laissez-faire 11, 36–37, 42
Language games 49–51
Late capitalism 1
Latour, Bruno 17, 167, 176–177
Law 9–10, 20–21, 35, 40, 50, 67, 88, 125, 131, 143, 147, 156, 165, 176, 181, 202, 211, 221
Lawyers 160
Lazzarato, Maurizio 13, 18, 138, 151, 155, 158
Le Corbusier 27, 43, 60–61, 68, 76, 91, 187, 190
Lean Production 149, 157–158
Lefort, Claude 9, 17
Legal standard 210
Leibniz, Gottfried Wilhelm von 215
Lenin, Vladimir 43, 91
Liberal democracy 189
Liberalism 10–12, 114, 189
Liberty 12, 43, 67
Linearity 172
Lloyd, Genevieve 62, 76, 223, 229
Lobbying, Corporate 193
Local authority 225–227
Local, The 133
Loewy, Raymond 125–126
Logic 10, 12, 15, 39–40, 49, 52–54, 60, 68, 74, 81, 83–84, 87–91, 139, 141–143, 179, 181–182, 191, 196–199, 205–206, 208, 217, 228
Logistics 82, 86
Logo (computer program) 98–99, 101–106, 115, 117, 119–120

Low latency 164
Luhmann, Niklas 8, 32–33, 35, 43, 50–51, 79, 88–89
Lynn, Greg 15, 186, 204, 223
Lyotard, Jean-François 13, 17, 49–53, 56, 70, 77, 139–140, 157

MAD Architects 71
Machine aesthetic 115
Madrid 168–170, 172–173, 176–177
Management, Risk 13, 17, 44, 108, 138, 140, 142, 154, 156–158, 178, 180, 186, 198–199, 202, 205, 208, 221, 226–227
Managerial 6
Mandela, Nelson 65
Manferdini, Elena 71, 74
Manufacturing 16, 149–150, 157, 200, 203, 207, 209
Mapping 4, 15, 32, 63, 165
Marazzi, Christian 13, 18, 138
Marginalization 191, 198
Marketing 8, 148, 183, 209
Marketization 12
Markets 11–12, 37, 80, 84, 87, 96, 108, 143, 153, 177, 184, 217, 225, 230
Marx, Karl 9, 42, 145–146, 156–157, 217
Marxian 87
Marxism 33, 188
Massachusetts Institute of Technology (MIT) 96
Massive scale (computation) 159, 191
Materialism 93, 146, 179, 222, 228
Materialist 4, 10, 127, 143, 213–214, 217, 224
Mathematics 5, 48, 50, 52–54, 56–57, 90, 101–102, 105, 119, 143, 154, 201, 203, 207, 215–216, 228
Matter, Inert 219
Maturana, Humberto 8, 16
Maya (computer program) 6, 31, 203
McKinsey 87
McLuhan, Marshall 8, 17
Means of Production 139, 210
Media Lab (MIT) 96, 106–107, 111–117, 120–121
Mediation 10; Digital Mediation 173
Megacities 34, 37, 39, 44
Megacity 192

Meikle, Jeffrey 136
Meillassoux, Quentin 217
Menges, Achim 215, 228
Mental health 144, 225
Merleau-Ponty, Maurice 52
Meta-structure 221
Metadata 201
Metaphors, Digital 221
Metaphysical 46–47, 89, 97, 116
Metaphysics 16, 47, 100
Metrics 182, 215
MiArmy 31
Micro-society 168, 172
Microsoft 91
Mies van der Rohe, Ludwig 187
Minimalism 69
Minsky, Marvin 8, 96, 101, 106–107, 120
MIT (Massachusetts Institute of Technology) 15, 17, 56, 78, 96, 101, 104, 107–108, 111, 113–114, 118–121, 158, 211, 228
Model; Analytical 203; Computer 95; Financial 6, 238
Modeling, Climate 209
Modeling, Virtual 5–6
Modernism 5, 23–24, 26, 33–34, 37, 39, 43, 47, 55, 69, 120, 175, 182, 187, 214
Modernity 51, 56, 187
Modes of existence 219
Modes of production 156, 197–198
Money transfer services 171
Moore's Law 202, 211
Morpho-ecological method 215
Morphogenesis 129
Morphology 6, 36, 143, 218
Mortgages 225–226
Mosque 171, 175
Mouffe, Chantal 9, 17, 190, 199
Mouride Brotherhood 168
Mouse (computer) 134, 202
Multiplicity 43, 54, 173, 191
Multivariable Design Optimization 207
Museum of Modern Art (MoMA, New York) 121, 128, 134, 136

Naess, Arne 220, 223, 229
Nanotechnology 86
Narcissism 197
National Science Foundation 101, 105
National Security Agency (NSA) 90–91
Nationalism 121
Natural Resources 190
Naturalism 11, 141, 146, 215
Nature 11, 17, 60, 97–98, 117, 127, 129, 133, 135, 144, 146, 157, 171, 189, 214–220, 222–225, 227, 229
Neg-entropy 40
Negarestani, Reza 93
Negotiation 21, 173
Negroponte, Nicholas 96, 107–108, 111–115, 118, 120–121, 205, 211
Neo-avant-garde 190
Neo-conservative 110
Neo-Darwinian 129, 132
Neo-Marxism 188
Neo-Pragmatism 142
Neoclassicism 195
Neoliberal 3–4, 12, 15, 37, 81, 87, 133, 142–143, 179, 181, 183, 190–192, 214, 219–220, 226, 228; Codes of production 220; Governance 12, 193, 220–221, 226; Political economy 190
Neoliberalism 11–15, 42, 87, 142, 189–191, 193, 195, 197, 199
Neuro-marketing 8
Neurosis 178
New York 15–18, 27, 44, 56–57, 71, 76, 78, 93, 109, 118–123, 128, 136–137, 167, 188, 193–194, 197, 211–212, 228–229
New York World's Fair (1939) 123
Nietzsche, Friedrich 187
Nieuwenhuys, Constant 190
Nike Inc. 158
Non-anthropocentric power 219
Non-anthropomorphic design 217
Non-hierarchical 115
Non-Human 4, 86, 214, 217–222
Non-linearity 46
Non-pathological relations 221
Non-standard computational design 217
Nouvel, Jean 64
Novelty 5, 83, 148
Nurbs 29, 143, 215

Object-oriented CAD 202–203
Object-oriented programming 98
Objective facticity 217, 220
Objectivity 45

Office for Naval Research 108
Oil 194, 196, 239
One Laptop per Child 115
Online platforms 171
Ontology 7–8, 12–13, 26, 40, 46, 48, 50, 52, 53, 57, 116, 214, 217–220
Operaismo 13, 138
Opportunity (capitalist) 112, 209–211
Optimization 88, 131–132, 134, 207, 211
Ordoliberalism 12
Organic processes 100, 217
Organicism 54, 96, 115
Organism 7–8, 100, 106, 114, 127, 129, 134–135, 221
Origination 143, 214
Orthographic Drawings 201
Otherness 176, 223–224
Otto, Frei 70, 156
Ownership 170, 198

Pai, Hyungmin 45
Panopticon 66–67, 76
Papert, Seymour 96–98, 101–106, 116–120, 122
Parameter 1–2, 5, 15–16, 32, 47–48, 52, 54–55, 77, 108, 123, 140–141, 147–148, 154, 172–173, 190, 196–197, 200–205, 207, 209, 211, 215
Parametric 2, 4–7, 12, 15–16, 20, 26, 29–32, 38–39, 43, 47–48, 55–56, 69–71, 77, 80–82, 84, 86–88, 90–92, 117, 123, 127, 132, 138–139, 141, 143, 149, 154–155, 175–179, 181–185, 187–188, 190, 195–196, 200, 202–211, 213–214, 217, 223
Parametricism 1–6, 8–10, 12, 14–15, 19–21, 23–27, 29–35, 37–43, 47, 50, 54–55, 69–70, 75, 77, 79–81, 83, 85–88, 90–92, 96, 117–118, 123, 125, 127, 129, 131, 133, 135, 137, 142–143, 157, 168, 175–177, 179, 181–182, 184, 186, 188, 190–193, 196–197, 202, 204, 206–207, 213–218, 220, 222, 228
Parametrics 4–6, 15, 77, 89, 140, 143, 153–155, 178–180, 182–183, 187, 205–206, 215
Paris 27, 65, 168–170
Parsons, Talcott 51
Participatory design 108

Particle-Spring Systems 29
Patriarchal relationship 206
Patriot Act 165
Peirce, Charles Sanders 49
Pérez Gómez, Alberto 222, 228
Performance, Technical 201
Performativity 46, 50, 106, 171
Peripheral (computing) 103, 105, 177
Phenomenology 7, 10
Philanthropy 111
Photoshop 52
Photovoltaics 195
Phylogenesis 123–124, 136
Piaget, Jean 101–103
Picon, Antoine 215, 228, 240
Picturesque 196
Plan (project) 28, 38, 42, 66–67, 69, 81, 85, 91, 201–202, 230
Planetary-scale computation 80, 83, 85–87
Planning 20, 37–39, 44, 76, 84–85, 89, 107, 193, 199, 230
Planometric Projection 201
Platform 61, 80–81, 83–85, 87, 91, 159, 171, 195, 202–203, 206; Online 171
Plot 202
Plug-in 5
Pluralism 33, 47, 189
Poiesis 46
Police 63, 68, 170, 172–173, 177
Polis 55, 137
Political economy 9, 156, 190, 196–197
Political representation 198
Political speech 163
Political, The 9–10, 19–22, 24, 44, 49, 52–53, 60–64, 66–68, 80–81, 83, 86–87, 89–92, 94, 96, 98, 146, 171, 175, 191, 194–197, 199, 214, 225
Pollution 213
Popularity 2, 162
Post-9/11, 179
Post-Fordism 24, 139, 141–143, 150, 158
Post-Fordist 2, 5, 13, 19–20, 23–24, 29, 36–39, 43, 138, 141–142, 149–150, 152, 154, 156
Post-Taylorism 143
Post-Taylorist 138, 142, 154, 156
Postcapitalism 184
Posthumanism 4, 221

Postindustrial 13, 50–51, 138, 139, 143, 151, 153–154
Postmodernism 5, 24, 41, 56, 69, 76
Postmodernity 50, 70
Postpolitical 176
Poststructuralism 10, 33
Postwar 8, 56, 63, 107, 113, 193, 195
Potentiality 147
Poverty 110, 193, 230
Power relations 215, 217
Pragmatism/pragmatics 43, 46, 49, 51–53, 141–142, 168
Praxis 58, 66
Pre-digital 175
Pre-parametric 175
Preciado, Beatriz 223, 229
Privatization 11, 97, 110, 191–192
Production, Means of 139, 210
Pro Engineer (software) 203
Professional Liability Insurance 210
Profiling 173
Pro-forma, Developer's 197
Program (computer) 8–9, 11, 15, 32, 71, 91–92, 96, 98, 101–103, 105–108, 202, 207
Program (building) 34, 36, 42, 81, 87, 91–92
Programming (computer) 5, 15, 77, 96, 98–102, 105–106, 115, 166, 199, 204; Object-oriented 98
Project Collaboration 205
Project Delivery 210
Property, Private 12, 154
Proportions 125, 130, 185
ProPublica 163, 166
Protest 95, 133, 163–164
Protocol 2, 8, 88, 160, 195, 197–199, 205
Pseudoscience 131
Psychoanalysis 56, 181–182, 187
Psychology 8, 12, 47, 96, 101, 178; Behavioural 8, 12
Public, The 22–23, 72, 97, 179, 191–193
Python 160–161, 166

Race 125, 200
Radical Democracy 17, 51
Rahim, Ali 136
Rancière, Jacques 10, 17, 221
Ratio 45, 83, 213–215, 217–229

Rational 39, 52, 85, 189, 214, 218–219, 224
Rationalization 5, 84–85, 87, 222; Formal 5
Readymades 53
Reagan, Ronald 11, 109–110, 116
Real Estate 8, 36, 44, 60, 92, 163, 177
Recursion 6, 105–106
Reformism 61, 81, 109–110, 209
Reichstag, The 63
Relations of Distribution 149–150, 152–153, 156–157
Relations of Production 141, 148–150, 152–153, 156–157
Relativism 47, 191
Renaissance 24–25, 45, 63
Rendering 204
Repetition 30, 42, 53, 55, 98, 142, 145, 188, 216, 218
Representation, Digital 204; Political 198
Representative Democracy 21
Reprogramming 171
Research Laboratory of Electronics (MIT) 101
Resistance 67, 183–184
Resource depletion 213, 216
Resources 23, 34, 36, 39, 41, 60, 106, 137, 147, 173, 190, 192–193, 197–199, 206, 217, 227
Revit (computer program) 183
Revolution 2, 21, 37, 54, 60–61, 97, 116, 158, 182, 229
Reynolds, Craig W. 94–96, 98–100, 106, 118–119
Rhino (computer program) 177, 203, 205, 214
RIBA (Royal Institute of British Architects) 187, 230
Riefenstahl, Leni 92
Rifkin, Adrian 64
Rights 12, 27, 110, 125, 159, 197, 199, 223, 226
Rights, Inalienable 12
Risk management 13, 17, 44, 108, 138, 140, 142, 154, 156–158, 178, 180, 186, 198–199, 202, 205, 208, 221, 226–227
Robinson, Kim Stanley 95
Robot 86, 89, 71, 77, 96, 98, 102, 103, 105, 108, 119, 134, 137
Romanticized, Parametricism 214

Rossi, Aldo 62
Royal Institute of British Architects (RIBA) 230
Rules of Thumb 143, 201, 206
Ruskin, John 157, 215, 223, 228
Russell, Bertrand 48, 50, 56

San Diego 82, 194
San Francisco 136, 200
Scale 12, 34, 69, 82, 86, 88–90, 127–129, 134, 159, 164–165, 175, 201–202, 219–222, 226; Full-, 201–202; Massive 159, 191; Nested 11; Planetary 80, 83, 85–87
Scenario 7, 20, 30–31, 173, 214
Scepticism 216
Schizoanalysis 182–183
Schizophrenia 178–179, 181–183, 185–187, 228
Schmitt, Carl 9
Schumacher, Patrik 1, 4–6, 8, 14–15, 17, 19, 43, 56, 69–71, 77, 79–83, 85, 87–93, 125, 141–143, 157, 177, 186, 188, 190, 214–215, 228
Science 5, 8–12, 21, 23–24, 35, 51, 89, 110, 132, 137, 158, 187, 196, 217, 222, 225; Biophysical 218; Cognitive 8, 96, 101; Climate 47; Computer 47, 85, 96, 101, 131; Ecology 221; Economic 11; Social 8–9, 33; World 23
Science and Technology 10, 217
Science and Technology Studies 10
Science fiction 107, 137
Scientificism 11
Script 29, 32, 118, 129, 132, 143, 160–161, 164–166, 177, 203–206, 215, 218–220, 224; Computational 219
Scripting 5, 15–16, 29, 69–70, 138–139, 143, 166, 177–179, 207; Form-driven 178
Scully, Vincent 62, 76
Search Engine 74, 164
Search Functionality 163
Section (drawing) 66, 70, 201
Security 10, 17, 96, 108, 166, 173
Seduction 213
Self-determination 221, 223, 226
Self-determined life 221
Self-evolving 214

Self-improvement 209
Self-interest 98, 115, 117
Self-organization 51, 100, 115
Self-preservation 117
Self-same replication 214
Semiology 32, 43
Semiotics 10, 13, 49, 221
Senegal 168
Sense-reason 222
Sensitivity 31
Sensor 96, 133, 200, 209
Set Theory 48–49, 52–53
Sex 173, 223–225
Sexed technicity 224
Sexual difference 223
Sexual reproduction 223
Shannon, Claude 8, 17
Shelter (UK charity) 225, 230
Siegert, Bernhard 56
Sign 54, 149, 153, 156, 164–165, 167, 175, 178
Signalization 7
Silicon Valley 44, 200
Simmel, Georg 45, 56
Simulation 7, 12, 94–98, 100, 106, 127, 129, 200, 204–211; Behavioural 204, 210; Sociological 100, 210
Simulation Algorithms (SAs) 207
Sina Weibo 159–160, 165–166
Situationist International 53
Slum 191
Smith, Adam 9, 15, 144, 157, 187–188
Social Demarcation 176
Social Democracy 10, 69, 192, 197
Social Forms 6, 82, 220
Social Imaginary 51
Social Justice 8, 193
Social Market Economics 12
Social Networks 2, 198–199
Socialism 24, 51
Socialisme ou Barbarie 51
Socialization 105, 198
Societies of Control 152–155, 158
Sociology 9, 22, 96–97, 118, 176
Socrates 144
Software engineer 7
Software, Proprietary 5, 184
Solidarity 165
South Africa House (London) 65

Sovereignty 15, 80, 84–85, 117, 154, 158
Spain 72, 134, 170
Spatial calculation 172
Special Economic Zones 82
Specialization 19, 196
Spectacle 47, 97, 117, 185
Speculative realist philosophies 217
Spinoza, Baruch 213–215, 217–225, 227–229
Spivak, Gayatri 9
Splines 29, 143, 215
Spreadsheets 195, 207
Spufford, Francis 85
Spuybroek, Lars 215, 223, 228–229
Srnicek, Nick 87, 93
Stalin, Joseph 43
State (nation) 11–12, 37–39, 60, 64, 69, 80, 83–86, 91, 113, 116–117, 119, 122, 140, 144, 152, 183–184, 193
Statistics 5, 11, 47, 55, 114, 122, 166
Stereotype 30–31, 178–179, 187
Stern, Robert 203
Stiny, George 8, 17
Structural Computation Programs 202
Structure (engineering) 4, 13, 60, 102, 117, 185–186, 201
Style 2, 5–6, 14, 23–24, 26, 29, 32–34, 37, 39–41, 47, 53, 55, 59, 62, 68–71, 73, 77, 87, 140, 142, 195, 213–214; Epochal 5, 23, 32, 34, 37, 40–41; International 5, 195
Subdivs 29, 215
Subject, The modern 218
Subjectivity 12, 97, 115, 150–152, 156, 158, 214, 218, 220–221, 224, 226
Subjectivization 106, 115
Subjects-in-process 219
Sublime 95, 97, 106, 113, 118, 217
Substance 58, 80, 102, 219–223
Suburban sprawl 196
Superstructure 182, 188
Surplus 113, 195, 198
Surveillance 2, 76, 91, 164
Sustainability 199, 216, 222, 225–226
Swarm 5, 94, 96–99, 101, 112–113, 115–118
Systems Theory 8, 32–33, 79

Taboo 29–30, 142, 160
Taxation 192, 226

Taylor, Frederick Winslow 156
Taylorism 143, 156
Taylorist 13, 138, 142, 154, 156
Taylorization 182
Technicity, Sexed 224
Techno-determinism 171
Techno-neutrality 171
Techno-social 103, 120, 173
Techno-utopia 13–14, 242, 244
TED (Technology, Entertainment, Design) 114–115, 121
Telematics 140
Telemetry 200, 211
Teleological 5, 182, 218–219
Teleology 182, 184, 188
Temporality 90
Terraforming 86
Terror 51
Thatcher, Margaret 11
Things 10, 45, 48–49, 58, 67, 81–82, 95, 102, 105, 137, 139, 141, 177, 201, 205, 215
Thompson, D'Arcy Wentworth 187, 218, 228
Tijuana 82, 194
Time-shared computer 105
Top-down 20, 36, 123, 191–192, 196–199
Topology 92, 155
TOR 90–91
Totalitarianism 17, 21–22, 92
Tourism 63
Toyotism 150, 157
Transcendental laws of reason 221
Transnational 171–172
Trial-and-error 173
Trickle-down economics 192
Truth 11, 46–47, 49–50, 53, 100
Turing, Alan 8, 100, 119
Turtle Geometry 96, 98, 102–103
Twitter 159–160, 164
Typology 7, 62
Tyranny 47, 55

Umwelt 221, 224
Uncertainty 33, 69, 164
Uncommon, The 223
Unemployment 113
Unions, Workers' 11, 90

United Arab Emirates 194
Universal grammar 216
Universal wage 186
Universalism 224
Universalist 80–81, 83, 85, 87, 191
University of Hong Kong 160
Urban 14, 20, 29, 31, 35–42, 44, 58, 69, 82–83, 85–86, 88–90, 108, 110, 125, 128, 168, 171, 173, 175–177, 191–199, 215; Asymmetry 190–191; Conflicts 190–191, 195; Crisis 110, 191; Design 8, 26; 36, 69, 90, 192, 195; Development 20, 33, 36–39, 42, 133, 193–195, 198–199, 225; Discontent 108; Fields 5, 141; Growth 190, 195; Planning 37, 76, 89; Unrest 60, 62
Urban enactments 168, 175–176
Urbanism 20, 32, 34, 37–39, 42, 86, 125, 141–142, 169–171, 173–174, 195
Urbanism, Trans-border 194
Urbanization 20, 38, 42, 127, 190–191, 194–195, 197–199
User 5, 7, 12, 38–39, 60, 65, 73, 82–84, 86, 108, 130, 160, 163–164, 166
User account 160
Username 162
Utopia 14

Valentine, Maggie 62, 76
Varela, Francisco 8, 16, 51–52, 57
Vector, Digital 38, 42, 98, 196, 202; mathematical 202
Violence 18, 51, 191–192, 221
Virno, Paolo 13, 18, 138
Virtual Private Network (VPN) 159

Virtualization 209
Visual Data Set 160
Vitalism 222
Volume 123, 175, 196
von Uexküll, Jakob 221

Wage 186
Wall Street Journal 113–114
Warrant Canary 164–165
Wealth 14, 144, 146, 157, 193–194
Weather 208
WeiboScope 160, 162, 166
Welfare State 191
Wellbeing 214, 220, 225–227, 230
Western civilization 5, 87
Wholesale facilities 170
Wiener, Norbert 17
Wiesner, Jerry 107, 111, 113, 120
Wiesner building (Media Lab) 107, 111
Williams, Alex 87, 93, 230
Wittgenstein, Ludwig 48–53, 56, 68
Wodiczko, Krysztof 65
Wright, Frank Lloyd 62, 76

Xu Weiguo, Professor 71

Yansong Ma (architect) 71

Zaha Hadid 1, 14, 26, 28, 34, 43, 70, 125, 186
Zaha Hadid Architects 1, 14, 26, 28, 34, 43, 70
Zoe (philosophy) 220–221, 223
Zoning 30, 198
Zoo 113
Zurr, Ionat 128, 134–135, 137